Nanotechnology: Basic Science and Emerging Technologies

Nanotechnology: Basic Science and Emerging Technologies

By

Rohit Majumdar

2008

CYBER TECH PUBLICATIONS
4264/3, Ansari Road, Daryaganj, New Delhi-110002 (India)
Ph.: 011-23244078. Fax: 011-23280028
E-Mail: cyberpublicationsdelhi@yahoo.com

NANOTECHNOLOGY: BASIC SCIENCE AND EMERGING TECHNOLOGIES

Rohit Majumdar

First Edition 2008

Published by :

G.S. Rawat for **Cyber Tech Publications**
4264/3, Ansari Road, Daryaganj, New Delhi-110002 (India)
Ph.: 011-23244078. Fax: 011-23280028
E-Mail: cyberpublicationsdelhi@yahoo.com

Printed at : **Amit Enterprises**, Maujpur, Delhi, (9810878184)

Preface

Compared with physics, chemistry, and other physical sciences, nanotechnology has not drawn as consistently, as quickly, or as thoroughly on the capabilities afforded by emerging technologies. With some notable exceptions, traditional imaging, data collection, and analysis techniques have remained the foundation for progress in the biological sciences. However, many of the seminal advances in nanotechnology in recent years have been based in large part on the creative application of new technologies. It is likely that this trend will continue, as technological advances continue to open up opportunities for scientific advance.

The emerging fields of nanoscale science, engineering, and technology the ability to work at the molecular level, atom by atom, to create large structures with fundamentally new properties and functions are leading to unprecedented understanding and control over the basic building blocks and properties of all natural and man-made things. The objective of this book is to provide a knowledge of new developments, recent progress, and innovations in the design and implementation of MEMS, NANO, and Smart Systems-on-Chip. It covers all aspects of design methods of those systems. The emphasis is on current and future challenges in research and development in both academia and industry. It emphasizes long-term, fundamental research aimed at discovering novel phenomena, processes, and tools; addressing the integration of Nano, MEMS and Smart Systems grand challenges; supporting new interdisciplinary research activities and addressing research and educational activities on the societal implications of advances in nanoscience and nanotechnology.

I sincerely hope that this work of mine would be informative and pleasurable to both teachers and students alike. Suggestions and comments of the readers are always welcome.

—Author

Contents

1

INTRODUCTION

Truly revolutionary nanotechnology products, materials and applications, such as nanorobotics, are years in the future (some say only a few years; some say many years). What qualifies as "nanotechnology" today is basic research and development that is happening in laboratories all over the world. "Nanotechnology" products that are on the market today are mostly gradually improved products (using evolutionary nanotechnology) where some form of nanotechnology enabled material (such as carbon nanotubes, nanocomposite structures or nanoparticles of a particular substance) or nanotechnology process (e.g. nanopatterning or quantum dots for medical imaging) is used in the manufacturing process.

In their ongoing quest to improve existing products by creating smaller components and better performance materials, all at a lower cost, the number of companies that will manufacture "nanoproducts" (by this definition) will grow very fast and soon make up the majority of all companies across many industries. Evolutionary nanotechnology should therefore be viewed as a process that gradually will affect most companies and industries.

DEFINITION

So what exactly is nanotechnology? One of the problems facing nanotechnology is the confusion about its definition. Most definitions revolve around the study and control of phenomena and materials at length scales below 100 nm and quite often they make a comparison with a human hair, which is about 80,000 nm wide. Some definitions include a reference to molecular systems and devices and nanotechnology 'purists' argue that any definition of nanotechnology needs to include a reference to "functional systems". The inaugural issue of Nature Nanotechnology asked 13 researchers from different areas what nanotechnology means to them and the responses, from enthusiastic to sceptical, reflect a variety of perspectives.

It seems that a size limitation of nanotechnology to the 1-100 nm range, the area where size-dependant quantum effects come to bear, would exclude numerous materials and devices, especially in the pharamaceutical area, and some experts caution against a rigid definition based on a sub-100 nm size. We found a good definition that is practical and unconstrained by any arbitrary size limitations (source): The design, characterization, production, and application of structures, devices, and systems by controlled manipulation of size and shape at the nanometer scale (atomic, molecular, and macromolecular scale) that produces structures, devices, and systems with at least one novel/superior characteristic or property.

THE SIGNIFICANCE

A nanometre (nm) is one thousand millionth of a metre. For comparison, a red blood cell is approximately 7,000 nm wide and a water molecule is almost 0.3nm across. People are interested in the nanoscale (which we define to be from 100nm down to the size of atoms (approximately 0.2nm)) because it is at this scale that the properties of materials can be very different from those at a larger scale. We define nanoscience as the study of phenomena and manipulation of materials at atomic, molecular and macromolecular scales, where properties differ significantly from those at a larger scale; and nanotechnologies as the design, characterisation, production and application of structures, devices and systems by controlling shape and size at the nanometre scale. In some senses, nanoscience and nanotechnologies are not new. Chemists have been making polymers, which are large molecules made up of nanoscale subunits, for many decades and nanotechnologies have been used to create the tiny features on computer chips for the past 20 years. However, advances in the tools that now allow atoms and molecules to be examined and probed with great precision have enabled the expansion and development of nanoscience and nanotechnologies.

The properties of materials can be different at the nanoscale for two main reasons. First, nanomaterials have a relatively larger surface area when compared to the same mass of material produced in a larger form. This can make materials more chemically reactive (in some cases materials that are inert in their larger form are reactive when produced in their nanoscale form), and affect their strength or electrical properties. Second, quantum effects can begin to dominate the behaviour of matter at the nanoscale- particularly at the lower end- affecting the optical, electrical and magnetic behaviour of materials. Materials can be produced that are nanoscale in one dimension (for example, very thin surface coatings), in two dimensions (for example, nanowires and nanotubes) or in all three dimensions (for example, nanoparticles).

NEW MATERIALS

Much of nanoscience and many nanotechnologies are concerned with producing new or enhanced materials.

Nanomaterials can be constructed by 'top down' techniques, producing very small structures from larger pieces of material, for example by etching to create circuits on the surface of a silicon microchip. They may also be constructed by 'bottom up' techniques, atom by atom or molecule by molecule. One way of doing this is self-assembly, in which the atoms or molecules arrange themselves into a structure due to their natural properties.

Crystals grown for the semiconductor industry provide an example of self assembly, as does chemical synthesis of large molecules. A second way is to use tools to move each atom or molecule individually. Although this 'positional assembly' offers greater control over construction, it is currently very laborious and not suitable for industrial applications. Current applications of nanoscale materials include very thin coatings used, for example, in electronics and active surfaces (for example, self-cleaning windows). In most applications the nanoscale components will be fixed or embedded but in some, such as those used in cosmetics and in some pilot environmental remediation applications, free nanoparticles are used. The ability to machine materials to very high precision and accuracy (better than 100nm) is leading to considerable benefits in a wide range of industrial sectors, for example in the production of components for the information and communication technology, automotive and aerospace industries.

Definition

Although a broad definition, we categorise nanomaterials as those which have structured components with at least one dimension less than 100nm. Materials that have one dimension in the nanoscale (and are extended in the other two dimensions) are layers, such as a thin films or surface coatings. Some of the features on computer chips come in this category. Materials that are nanoscale in two dimensions (and extended in one dimension) include nanowires and nanotubes. Materials that are nanoscale in three dimensions are particles, for example precipitates, colloids and quantum dots (tiny particles of semiconductor materials). Nanocrystalline materials, made up of nanometre-sized grains, also fall into this category. Some of these materials have been available for some time; others are genuinely new. The aim of this chapter is to give an overview of the properties, and the significant foreseeable applications of some key nanomaterials.

Two principal factors cause the properties of nanomaterials to differ significantly from other materials: increased relative surface area, and quantum effects. These factors can change or enhance properties such as reactivity, strength and electrical characteristics. As a particle decreases in size, a greater proportion of atoms are found at the surface compared to those inside. For example, a particle of size 30 nm has 5% of its atoms on its surface, at 10 nm 20% of its atoms, and at 3 nm 50% of its atoms. Thus nanoparticles have a much greater surface area per unit mass compared with larger particles. As growth and catalytic chemical reactions occur at surfaces, this means that a given mass of material in nanoparticulate form will be much more reactive than the same mass of material made up of larger particles.

To understand the effect of particle size on surface area, consider a U.S. silver dollar. The silver dollar contains 26.96 grams of coin silver, has a diameter of about 40 mm, and has a total surface area of approximately 27.70 square centimeters. If the same amount of coin silver were divided into tiny particles - say 1 nanometer in diameter - the total surface area of those particles would be 11,400 square meters. When the amount of coin silver contained in a silver dollar is rendered into 1 nm particles, the surface area of those particles is 4.115 million times greater than the surface area of the silver dollar!

Properties

In tandem with surface-area effects, quantum effects can begin to dominate the properties of matter as size is reduced to the nanoscale. These can affect the optical, electrical and magnetic behaviour of materials, particularly as the structure or particle size approaches the smaller end of the nanoscale. Materials that exploit these effects include quantum dots, and quantum well lasers for optoelectronics.

For other materials such as crystalline solids, as the size of their structural components decreases, there is much greater interface area within the material; this can greatly affect both mechanical and electrical properties. For example, most metals are made up of small crystalline grains; the boundaries between the grain slow down or arrest the propagation of defects when the material is stressed, thus giving it strength. If these grains can be made very small, or even nanoscale in size, the interface area within the material greatly increases, which enhances its strength. For example, nanocrystalline nickel is as strong as hardened steel. Understanding surfaces and interfaces is a key challenge for those working on nanomaterials, and one where new imaging and analysis instruments are vital.

SCIENCE OF NANOMATERIAL

Nanomaterials are not simply another step in the miniaturization of materials. They often require very different production approaches. There are several processes to create nanomaterials, classified as 'top-down' and 'bottom-up'. Although many nanomaterials are currently at the laboratory stage of manufacture, a few of them are being commercialised. Below we outline some examples of nanomaterials and the range of nanoscience that is aimed at understanding their properties. As will be seen, the behaviour of some nanomaterials is well understood, whereas others present greater challenges.

Nanoscale In One Dimension

Thin Films, Layers and Surfaces

One-dimensional nanomaterials, such as thin films and engineered surfaces, have been developed and used for decades in fields such as electronic device manufacture, chemistry and engineering. In the silicon integrated-circuit industry, for example, many devices rely on thin films for their operation, and control of film thicknesses approaching the atomic level is routine. Monolayers (layers that are one atom or molecule deep) are also routinely made and used in chemistry. The formation and properties of these layers are reasonably well understood from the atomic level upwards, even in quite complex layers (such as lubricants). Advances are being made in the control of the composition and smoothness of surfaces, and the growth of films. Engineered surfaces with tailored properties such as large surface area or specific reactivity are used routinely in a range of applications such as in fuel cells and catalysts. The large surface area provided by nanoparticles, together with their ability to self assemble on a support surface, could be of use in all of these applications.

Although they represent incremental developments, surfaces with enhanced properties should find applications throughout the chemicals and energy sectors. The benefits could surpass the obvious economic and resource savings achieved by higher activity and greater selectivity in reactors and separation processes, to enabling small-scale distributed processing (making chemicals as close as possible to the point of use). There is already a move in the chemical industry towards this. Another use could be the small-scale, on-site production of high value chemicals such as pharmaceuticals.

Two Dimensions of Nanoscale

Two dimensional nanomaterials such as tubes and wires have generated considerable interest among the scientific community in recent years. In particular, their novel electrical and mechanical properties are the subject of intense research.

Carbon Nanotubes

Carbon nanotubes (CNTs) were first observed by Sumio Iijima in 1991. CNTs are extended tubes of rolled graphene sheets. There are two types of CNT: single-walled (one tube) or multi-walled (several concentric tubes). Both of these are typically a few nanometres in diameter and several micrometres to centimetres long. CNTs have assumed an important role in the context of nanomaterials, because of their novel chemical and physical properties. They are mechanically very strong (their Young's modulus is over 1 terapascal, making CNTs as stiff as diamond), flexible (about their axis), and can conduct electricity extremely well (the helicity of the graphene sheet determines whether the CNT is a semiconductor or metallic). All of these remarkable properties give CNTs a range of potential applications: for example, in reinforced composites, sensors, nanoelectronics and display devices.

Models of different singlewall nanotubes

CNTs are now available commercially in limited quantities. They can be grown by several techniques.

However, the selective and uniform production of CNTs with specific dimensions and physical properties is yet to be achieved. The potential similarity in size and shape between CNTs and asbestos fibres has led to concerns about their safety.

Inorganic Nanotubes

Inorganic nanotubes and inorganic fullerene-like materials based on layered compounds such as molybdenum disulphide were discovered shortly after CNTs. They have excellent tribological (lubricating) properties, resistance to shockwave impact, catalytic reactivity, and high capacity for hydrogen and lithium storage, which suggest a range of promising applications. Oxide-based nanotubes (such as titanium dioxide) are being explored for their applications in catalysis, photo-catalysis and energy storage.

Nanowires

Nanowires are ultrafine wires or linear arrays of dots, formed by self-assembly. They can be made from a wide range of materials. Semiconductor nanowires made of silicon, gallium nitride and indium phosphide have demonstrated remarkable optical, electronic and magnetic characteristics (for example, silica nanowires can bend light around very tight corners). Nanowires have potential applications in high-density data storage, either as magnetic read heads or as patterned storage media, and electronic and opto-electronic nanodevices, for metallic interconnects of quantum devices and nanodevices.

The preparation of these nanowires relies on sophisticated growth techniques, which include selfassembly processes, where atoms arrange themselves naturally on stepped surfaces, chemical vapour deposition (CVD) onto patterned substrates, electroplating or molecular beam epitaxy (MBE). The 'molecular beams' are typically from thermally evaporated elemental sources.

Biopolymers

The variability and site recognition of biopolymers, such as DNA molecules, offer a wide range of opportunities for the self-organization of wire nanostructures into much more complex patterns. The DNA backbones may then, for example, be coated in metal. They also offer opportunities to link nano- and biotechnology in, for example, biocompatible sensors and small, simple motors.

Such self-assembly of organic backbone nanostructures is often controlled by weak interactions, such as hydrogen bonds, hydrophobic, or van der Waals interactions (generally in aqueous environments) and hence requires quite different synthesis strategies to CNTs, for example. The combination of one-dimensional nanostructures consisting of biopolymers and inorganic compounds opens up a number of scientific and technological opportunities.

Three Dimensions of Nanoscale

Nanoparticles

Nanoparticles are often defined as particles of less than 100nm in diameter. We classify nanoparticles to be particles less than 100nm in diameter that exhibit new or enhanced size-dependent properties compared with larger particles of the same material. Nanoparticles exist widely in the natural world: for example as the products of photochemical and volcanic activity, and created by plants and algae. They have also been created for thousands of years as products of combustion and food cooking, and more recently from vehicle exhausts. Deliberately manufactured nanoparticles, such as metal oxides, are by comparison in the minority. Nanoparticles are of interest because of the new properties (such as chemical reactivity and optical behaviour) that they exhibit compared with larger particles of the same materials. For example, titanium dioxide and zinc oxide become transparent at the nanoscale, however are able to absorb and reflect UV light, and have found application in sunscreens. Nanoparticles have a range of potential applications: in the short-term in new cosmetics, textiles and paints; in the longer term, in methods of targeted drug delivery where they could be to used deliver drugs to a specific site in the body.

Nanoparticles can also be arranged into layers on surfaces, providing a large surface area and hence enhanced activity, relevant to a range of potential applications such as catalysts. Manufactured nanoparticles are typically not products in their own right, but generally serve as raw materials, ingredients or additives in existing products. Nanoparticles are currently in a small number of consumer products such as cosmetics and their enhanced or novel properties may have implications for their toxicity. For most applications, nanoparticles will be fixed (for example, attached to a surface or within in a composite) although in others they will be free or suspended in fluid. Whether they are fixed or free will have a significant affect on their potential health, safety and environmental impacts.

Fullerenes (carbon 60)

In the mid-1980s a new class of carbon material was discovered called carbon 60 (C60).Harry Kroto and Richard Smalley, the experimental chemists who discovered C60 named it "buckminsterfullerene", in recognition of the architect Buckminster Fuller, who was well-known for building geodesic domes, and the term fullerenes was then given to any closed carbon cage. C60 are spherical molecules about 1nm in diameter, comprising 60 carbon atoms arranged as 20 hexagons and 12 pentagons: the configuration of a football. In 1990, a technique to produce larger quantities of C60 was developed by resistively heating graphite rods in a helium atmosphere. Several applications are envisaged for fullerenes, such as miniature 'ball bearings' to lubricate surfaces, drug delivery vehicles and in electronic circuits.

Dendrimers

Dendrimers are spherical polymeric molecules, formed through a nanoscale hierarchical self-assembly process. There are many types of dendrimer; the smallest is several nanometres in size. Dendrimers are used in conventional applications such as coatings and inks, but they also have a range of interesting properties which could lead to useful applications. For example, dendrimers can act as nanoscale carrier molecules and as such could be used in drug delivery. Environmental clean-up could be assisted by dendrimers as they can trap metal ions, which could then be filtered out of water with ultra-filtration techniques.

Quantum Dots

Nanoparticles of semiconductors (quantum dots) were theorized in the 1970s and initially created in the early 1980s.

If semiconductor particles are made small enough, quantum effects come into play, which limit the energies at which electrons and holes (the absence of an electron) can exist in the particles. As energy is related to wavelength (or colour), this means that the optical properties of the particle can be finely tuned depending on its size. Thus, particles can be made to emit or absorb specific wavelengths (colours) of light, merely by controlling their size.

Recently, quantum dots have found applications in composites, solar cells (Gratzel cells) and fluorescent biological labels (for example to trace a biological molecule) which use both the small particle size and tuneable energy levels. Recent advances in chemistry have resulted in the preparation of monolayer-protected, high-quality, monodispersed, crystalline quantum dots as small as 2nm in diameter, which can be conveniently treated and processed as a typical chemical reagent.

APPLICATIONS

Below we list some key current and potential shortand long-term applications of nanomaterials. Most current applications represent evolutionary developments of existing technologies: for example, the reduction in size of electronics devices.

Current Applications

Sunscreens and Cosmetics

Nanosized titanium dioxide and zinc oxide are currently used in some sunscreens, as they absorb and reflect ultraviolet (UV) rays and yet are transparent to visible light and so are more appealing to the consumer. Nanosized iron oxide is present in some lipsticks as a pigment but it is our understanding that it is not used by the European cosmetics sector. The use of nanoparticles in cosmetics has raised a number of concerns about consumer safety.

Composites

An important use of nanoparticles and nanotubes is in composites, materials that combine one or more separate components and which are designed to exhibit overall the best properties of each component. This multi-functionality applies not only to mechanical properties, but extends to optical, electrical and magnetic ones. Currently, carbon fibres and bundles of multi-walled CNTs are used in polymers to control or enhance conductivity, with applications such as antistatic packaging.

The use of individual CNTs in composites is a potential long-term application. A particular type of nanocomposite is where nanoparticles act as fillers in a matrix; for example, carbon black used as a filler to reinforce car tyres. However, particles of carbon black can range from tens to hundreds of nanometres in size, so not all carbon black falls within our definition of nanoparticles.

Clays

Clays containing naturally occurring nanoparticles have long been important as construction materials and are undergoing continuous improvement. Clay particle based composites - containing plastics and nano-sized flakes of clay - are also finding applications such as use in car bumpers.

Coatings and Surfaces

Coatings with thickness controlled at the nano- or atomic scale have been in routine production for some time, for example in molecular beam epitaxy or metal oxide chemical vapor depositionfor optoelectonic devices, or in catalytically active and chemically functionalized surfaces. Recently developed applications include the self-cleaning window, which is coated in highly activated titanium dioxide, engineered to be highly hydrophobic (water repellent) and antibacterial, and coatings based on nanoparticulate oxides that catalytically destroy chemical agents.

Wear and scratch-resistant hard coatings are significantly improved by nanoscale intermediate layers (or multilayers) between the hard outer layer and the substrate material. The intermediate layers give good bonding and graded matching of elastic and thermal properties, thus improving adhesion. A range of enhanced textiles, such as breathable, waterproof and stainresistant fabrics, have been enabled by the improved control of porosity at the nanoscale and surface roughness in a variety of polymers and inorganics.

Tougher and Harder Cutting Tools

Cutting tools made of nanocrystalline materials, such as tungsten carbide, tantalum carbide and titanium carbide, are more wear and erosion-resistant, and last longer than their conventional (large-grained) counterparts. They are finding applications in the drills used to bore holes in circuit boards.

Short-term Applications

Paints

Incorporating nanoparticles in paints could improve their performance, for example by making them lighter and giving them different properties. Thinner paint coatings ('lightweighting'), used for example on aircraft, would reduce their weight, which could be beneficial to the environment. However, the whole life cycle of the aircraft needs to be considered before overall benefits can be claimed. It may also be possible to substantially reduce solvent content of paints, with resulting environmental benefits. New types of foulingresistant marine paint could be developed and are urgently needed as alternatives to tributyl tin (TBT), now that the ecological impacts of TBT have been recognised.

Anti-fouling surface treatment is also valuable in process applications such as heat exchange, where it could lead to energy savings. If they can be produced at sufficiently low cost, fouling-resistant coatings could be used in routine duties such as piping for domestic and industrial water systems. It remains speculation whether very effective anti-fouling coatings could reduce the use of biocides, including chlorine. Other novel, and more long-term, applications for nanoparticles might lie in paints that change colour in response to change in temperature or chemical environment, or paints that have reduced infra-red absorptivity and so reduce heat loss.

Concerns about the health and environmental impacts of nanoparticles may require the need for the durability and abrasion behaviour of nano-engineered paints and coatings to be addressed, so that abrasion products take the form of coarse or microscopic agglomerates rather than individual nanoparticles.

Remediation

The potential of nanoparticles to react with pollutants in soil and groundwater and transform them into harmless compounds is being researched. In one pilot study the large surface area and high surface reactivity of iron nanoparticles were exploited to transform chlorinated hydrocarbons (some of which are believed to be carcinogens) into less harmful end products in groundwater. It is also hoped that they could be used to transform heavy metals such as lead and mercury from bioavailable forms into insoluble forms. Serious concerns have been raised over the uncontrolled release of nanoparticles into the environment.

Fuel Cells

Engineered surfaces are essential in fuel cells, where the external surface properties and the pore structure affect performance. The hydrogen used as the immediate fuel in fuel cells may be generated from hydrocarbons by catalytic reforming, usually in a reactor module associated directly with the fuel cell. The potential use of nano-engineered membranes to intensify catalytic processes could enable higher-efficiency, small-scale fuel cells. These could act as distributed sources of electrical power. It may eventually be possible to produce hydrogen locally from sources other than hydrocarbons, which are the feedstocks of current attention.

Displays

The huge market for large area, high brightness, flat-panel displays, as used in television screens and computer monitors, is driving the development of some nanomaterials. Nanocrystalline zinc selenide, zinc sulphide, cadmium sulphide and lead telluride synthesized by sol-gel techniques (a process for making ceramic and glass materials, involving the transition from a liquid 'sol' phase to a solid 'gel' phase) are candidates for the next generation of light-emitting phosphors. CNTs are being investigated for low voltage field-emission displays; their strength, sharpness, conductivity and inertness make them potentially very efficient and long-lasting emitters.

Batteries

With the growth in portable electronic equipment (mobile phones, navigation devices, laptop computers, remote sensors), there is great demand for lightweight, high-energy density batteries. Nanocrystalline materials synthesized by sol-gel techniques are candidates for separator plates in batteries because of their foam-like (aerogel) structure, which can hold considerably more energy than conventional ones. Nickel-metal hydride batteries made of nanocrystalline nickel and metal hydrides are envisioned to require less frequent recharging and to last longer because of their large grain boundary (surface) area.

Fuel Additives

Research is underway into the addition of nanoparticulate ceria (cerium oxide) to diesel fuel to improve fuel economy by reducing the degradation of fuel consumption over time.

Catalysts

In general, nanoparticles have a high surface area, and hence provide higher catalytic activity. Nanotechnologies are enabling changes in the degree of control in the production of nanoparticles, and the support structure on which they reside. It is possible to synthesise metal nanoparticles in solution in the presence of a surfactant to form highly ordered monodisperse films of the catalyst nanoparticles on a surface. This allows more uniformity in the size and chemical structure of the catalyst, which in turn leads to greater catalytic activity and the production of fewer byproducts.

It may also be possible to engineer specific or selective activity. These more active and durable catalysts could find early application in cleaning up waste streams. This will be particularly beneficial if it reduces the demand for platinum-group metals, whose use in standard catalytic units is starting to emerge as a problem, given the limited availability of these metals.

LONGER-TERM APPLICATIONS

Carbon Nanotube Composites

CNTs have exceptional mechanical properties, particularly high tensile strength and light weight. An obvious area of application would be in nanotubereinforced composites, with performance beyond current carbon-fibre composites. One current limit to the introduction of CNTs in composites is the problem of structuring the tangle of nanotubes in a well-ordered manner so that use can be made of their strength. Another challenge is generating strong bonding between CNTs and the matrix, to give good overall composite performance and retention during wear or erosion of composites.

The surfaces of CNTs are smooth and relatively unreactive, and so tend to slip through the matrix when it is stressed. One approach that is being explored to prevent this slippage is the attachment of chemical side-groups to CNTs, effectively to form 'anchors'. Another limiting factor is the cost of production of CNTs. However, the potential benefits of such light, high strength material in numerous applications for transportation are such that significant further research is likely.

Lubricants

Nanospheres of inorganic materials could be used as lubricants, in essence by acting as nanosized 'ball bearings'. The controlled shape is claimed to make them more durable than conventional solid lubricants and wear additives. Whether the increased financial and resource cost of producing them is offset by the longer service life of lubricants and parts remains to be investigated. It is also claimed that these nanoparticles reduce friction between metal surfaces, particularly at high normal loads. If so, they should find their first applications in high-performance engines and drivers; this could include the energy sector as well as transport.

There is a further claim that this type of lubricant is effective even if the metal surfaces are not highly smooth. Again, the benefits of reduced cost and resource input for machining must be compared against production of nanolubricants. In all these applications, the particles would be dispersed in a conventional liquid lubricant; design of the lubricant system must therefore include measures to contain and manage waste.

Magnetic Materials

It has been shown that magnets made of nanocrystalline yttrium-samarium-cobalt grains possess unusual magnetic properties due to their extremely large grain interface area (high coercivity can be obtained because magnetization flips cannot easily propagate past the grain boundaries).

This could lead to applications in motors, analytical instruments like magnetic resonance imaging (MRI), used widely in hospitals, and microsensors. Overall magnetisation, however, is currently limited by the ability to align the grains' direction of magnetisation.

Nanoscale-fabricated magnetic materials also have applications in data storage. Devices such as computer hard disks depend on the ability to magnetize small areas of a spinning disk to record information. If the area required to record one piece of information can be shrunk in the nanoscale (and can be written and read reliably), the storage capacity of the disk can be improved dramatically.

In the future, the devices on computer chips which currently operate using flows of electrons could use the magnetic properties of these electrons, called spin, with numerous advantages. Recent advances in novel magnetic materials and their nanofabrication are encouraging in this respect.

Medical Implants

Current medical implants, such as orthopaedic implants and heart valves, are made of titanium and stainless steel alloys, primarily because they are biocompatible. Unfortunately, in some cases these metal alloys may wear out within the lifetime of the patient. Nanocrystalline zirconium oxide (zirconia) is hard, wearresistant, bio-corrosion resistant and bio-compatible. It therefore presents an attractive alternative material for implants.

It and other nanoceramics can also be made as strong, light aerogels by sol-gel techniques. Nanocrystalline silicon carbide is a candidate material for artificial heart valves primarily because of its low weight, high strength and inertness.

Machinable Ceramics

Ceramics are hard, brittle and difficult to machine. However, with a reduction in grain size to the nanoscale, ceramic ductility can be increased. Zirconia, normally a hard, brittle ceramic, has even been rendered superplastic (for example, able to be deformed up to 300% of its original length).

Nanocrystalline ceramics, such as silicon nitride and silicon carbide, have been used in such automotive applications as high-strength springs, ball bearings and valve lifters, because they can be easily formed and machined, as well as exhibiting excellent chemical and high-temperature properties. They are also used as components in high-temperature furnaces. Nanocrystalline ceramics can be pressed into complex net shapes and sintered at significantly lower temperatures than conventional ceramics.

Water Purification

Nano-engineered membranes could potentially lead to more energy-efficient water purification processes, notably in desalination by reverse osmosis. Again, these applications would represent incremental improvements in technologies that are already available. They would use fixed nanoparticles, and are therefore distinct from applications that propose to use free nanoparticles.

Military Battle Suits

Enhanced nanomaterials form the basis of a state-of-the-art 'battle suit' that is being developed by the Institute of Soldier Nanotechnologies at MIT. A short-term development is likely to be energy-absorbing materials that will withstand blast waves; longer-term are those that incorporate sensors to detect or respond to chemical and biological weapons (for example, responsive nanopores that 'close' upon detection of a biological agent). There is speculation that developments could include materials which monitor physiology while a soldier is still on the battlefield, and uniforms with potential medical applications, such as splints for broken bones.

2

NANOTECHNOLOGY AND CARBON NANOTUBES

The intent of this chapter is to convey a general understanding of what carbon Nanotubes are, how they are produced, their many unique and interesting properties, markets, and applications.

HISTORY

In 1980 we knew of only three forms of carbon, namely diamond, graphite, and amorphous carbon. Today we know there is a whole family of other forms of carbon. The first to be discovered was the hollow, cage-like buckminsterfullerene molecule- also known as the buckyball, or the C60 fullerene. There are now thirty or more forms of fullerenes, and also an extended family of linear molecules, carbon nanotubes. C60 is the first spherical carbon molecule, with carbon atoms arranged in a soccer ball shape. In the structure there are 60 carbon atoms and a number of five-membered rings isolated by six-membered rings.

The second, slightly elongated, spherical carbon molecule in the same group resembles a rugby ball, has seventy carbon atoms and is known as C70. C70's structure has extra six-membered carbon rings, but there are also a large number of other potential structures containing the same number of carbon atoms. Their particular shapes depend on whether five-membered rings are isolated or not, or whether seven-membered rings are present. Many other forms of fullerenes up to and beyond C120 have been characterized, and it is possible to make other fullerene structures with five-membered rings in different positions and sometimes adjoining one another.

The important fact for nanotechnology is that useful dopant atoms can be placed inside the hollow fullerene ball. Atoms contained within the fullerene are said to be endohedral. Of course they can also be bonded to fullerenes outside the ball as salts, if the fullerene can gain electrons.

Endohedral fullerenes can be produced in which metal atoms are captured within the fullerene cages. Theory shows that the maximum electrical conductivity is to be expected for endohedral metal atoms, which will transfer three electrons to the fullerene. Fullerenes can be dispersed on the surface as a monolayer. That is, there is only one layer of molecules, and they are said to be mono dispersed. Provided fullerenes can be placed in very specific locations, they may be aligned to form a fullerene wire.

Systems with appropriate material inside the fullerene ball are conducting and are of particular interest because they can be deposited to produce bead-like conducting circuits. Combining endohedrally doped structures with non-doped structures changes the actual composition of a fullerene wire, so that it may be tailored in-situ during patterning. Hence within a single wire, insulating and conducting regions may be precisely defined. One-dimensional junction engineering becomes realistic with fullerenes.

Possibly more important than fullerenes are Carbon nanotubes, which are related to graphite. The molecular structure of graphite resembles stacked, one-atom-thick sheets of chicken wire- a planar network of interconnected hexagonal rings of carbon atoms. In conventional graphite, the sheets of carbon are stacked on top of one another, allowing them to easily slide over each other.

That is why graphite is not hard, but it feels greasy, and can be used as a lubricant. When graphene sheets are rolled into a cylinder and their edges joined, they form CNTs. Only the tangents of the graphitic planes come into contact with each other, and hence their properties are more like those of a molecule.

CNTs come in a variety of diameters, lengths, and functional group content. CNTs today are available for industrial applications in bulk quantities up metric ton quantities from Cheap Tubes. Several CNT manufacturers have >100 ton per year production capacity for multi walled nanotubes.

A nanotube may consist of one tube of graphite, a one-atom thick single-wall nanotube, or a number of concentric tubes called multiwalled nanotubes. When viewed with a transmission electron microscope these tubes appear as planes. Whereas single walled nanotubes appear as two planes, in multi walled nanotubes more than two planes are observed, and can be seen as a series of parallel lines. There are different types of CNTs, because the graphitic sheets can be rolled in different ways. The three types of CNTs are Zigzag, Armchair, and Chiral. It is possible to recognize zigzag, armchair, and chiral CNTs just by following the pattern across the diameter of the tubes, and analyzing their cross-sectional structure.

Multi walled nanotubes can come in an even more complex array of forms, because each concentric single-walled nanotube can have different structures, and hence there are a variety of sequential arrangements. The simplest sequence is when concentric layers are identical but different in diameter. However, mixed variants are possible, consisting of two or more types of concentric CNTs arranged in different orders. These can have either regular layering or random layering. The structure of the nanotube influences its properties- including electrical and thermal conductivity, density, and lattice structure. Both type and diameter are important. The wider the diameter of the nanotube, the more it behaves like graphite. The narrower the diameter of the nanotube, the more its intrinsic properties depends upon its specific type.

PRODUCTION METHODS

There are a number of methods of making CNTs and fullerenes. Fullerenes were first observed after vaporizing graphite with a short-pulse, high-power laser, however this was not a practical method for making large quantities. CNTs have probably been around for a lot longer than was first realized, and may have been made during various carbon combustion and vapor deposition processes, but electron microscopy at that time was not advanced enough to distinguish them from other types of tubes. The first method for producing CNTs and fullerenes in reasonable quantities - was by applying an electric current across two carbonaceous electrodes in an inert gas atmosphere.

This method is called plasma arcing. It involves the evaporation of one electrode as cations followed by deposition at the other electrode. This plasma-based process is analogous to the more familiar electroplating process in a liquid medium. Fullerenes and CNTs are formed by plasma arcing of carbonaceous materials, particularly graphite. The fullerenes appear in the soot that is formed, while the CNTs are deposited on the opposing electrode. Another method of nanotube synthesis involves plasma arcing in the presence of cobalt with a 3% or greater concentration.

As noted above, the nanotube product is a compact cathode deposit of rod like morphology. However when cobalt is added as a catalyst, the nature of the product changes to a web, with strands of 1mm or so thickness that stretch from the cathode to the walls of the reaction vessel. The mechanism by which cobalt changes this process is unclear, however one possibility is that such metals affect the local electric fields and hence the formation of the five-membered rings.

Arc Method

The carbon arc discharge method, initially used for producing C60 fullerenes, is the most common and perhaps easiest way to produce CNTs, as it is rather simple. However, it is a technique that produces a complex mixture of components, and requires further purification- to separate the CNTs from the soot and the residual catalytic metals present in the crude product. This method creates CNTs through arc-vaporization of two carbon rods placed end to end, separated by approximately 1mm, in an enclosure that is usually filled with inert gas at low pressure.

Recent investigations have shown that it is also possible to create CNTs with the arc method in liquid nitrogen. A direct current of 50 to 100 A, driven by a potential difference of approximately 20 V, creates a high temperature discharge between the two electrodes. The discharge vaporizes the surface of one of the carbon electrodes, and forms a small rod-shaped deposit on the other electrode. Producing CNTs in high yield depends on the uniformity of the plasma arc, and the temperature of the deposit forming on the carbon electrode.

Laser Method

In 1996 CNTs were first synthesized using a dual-pulsed laser and achieved yields of >70wt% purity. Samples were prepared by laser vaporization of graphite rods with a 50:50 catalyst mixture of Cobalt and Nickel at 1200°C in flowing argon, followed by heat treatment in a vacuum at 1000°C to remove the C60 and other fullerenes. The initial laser vaporization pulse was followed by a second pulse, to vaporize the target more uniformly. The use of two successive laser pulses minimizes the amount of carbon deposited as soot.

The second laser pulse breaks up the larger particles ablated by the first one, and feeds them into the growing nanotube structure. The material produced by this method appears as a mat of "ropes", 10-20nm in diameter and up to 100μm or more in length. Each rope is found to consist primarily of a bundle of single walled nanotubes, aligned along a common axis.

By varying the growth temperature, the catalyst composition, and other process parameters, the average nanotube diameter and size distribution can be varied. Arc-discharge and laser vaporization are currently the principal methods for obtaining small quantities of high quality CNTs.

However, both methods suffer from drawbacks. The first is that both methods involve evaporating the carbon source, so it has been unclear how to scale up production to the industrial level using these approaches. The second issue relates to the fact that vaporization methods grow CNTs in highly tangled forms, mixed with unwanted forms of carbon and/or metal species. The CNTs thus produced are difficult to purify, manipulate, and assemble for building nanotube-device architectures for practical applications.

Chemical Vapor Deposition

Chemical vapor deposition of hydrocarbons over a metal catalyst is a classical method that has been used to produce various carbon materials such as carbon fibers and filaments. for over twenty years. Large amounts of CNTs can be formed by catalytic CVD of acetylene over Cobalt and iron catalysts supported on silica or zeolite. The carbon deposition activity seems to relate to the cobalt content of the catalyst, whereas the CNTs' selectivity seems to be a function of the pH in catalyst preparation.

Fullerenes and bundles of single walled nanotubes were also found among the multi walled nanotubes produced on the carbon/zeolite catalyst. Some researchers are experimenting with the formation of CNTs from ethylene. Supported catalysts such as iron, cobalt, and nickel, containing either a single metal or a mixture of metals, seem to induce the growth of isolated single walled nanotubes or single walled nanotubes bundles in the ethylene atmosphere. The production of single walled nanotubes, as well as double-walled CNTs, on molybdenum and molybdenum-iron alloy catalysts has also been demonstrated. CVD of carbon within the pores of a thin alumina template with or without a Nickel catalyst has been achieved. Ethylene was used with reaction temperatures of 545°C for Nickel-catalyzed CVD, and 900°C for an uncatalyzed process.

The resultant carbon nanostructures have open ends, with no caps. Methane has also been used as a carbon source. In particular it has been used to obtain 'nanotube chips' containing isolated single walled nanotubes at controlled locations. High yields of single walled nanotubes have been obtained by catalytic decomposition of an H2/ CH4 mixture over well-dispersed metal particles such as Cobalt, Nickel, and Iron on magnesium oxide at 1000°C.

It has been reported that the synthesis of composite powders containing well-dispersed CNTs can be achieved by selective reduction in an H2/CH4 atmosphere of oxide solid solutions between a non-reducible oxide such as Al2O3 or MgAl2O4 and one or more transition metal oxides. The reduction produces very small transition metal particles at a temperature of usually >800°C. The decomposition of CH4 over the freshly formed nanoparticles prevents their further growth, and thus results in a very high proportion of single walled nanotubes and fewer multi walled nanotubes.

Ball Milling

Ball milling and subsequent annealing is a simple method for the production of CNTs. Although it is well established that mechanical attrition of this type can lead to fully nano porous microstructures, it was not until a few years ago that CNTs of carbon and boron nitride were produced from these powders by thermal annealing. Essentially the method consists of placing graphite powder into a stainless steel container along with four hardened steel balls. The container is purged, and argon is introduced. The milling is carried out at room temperature for up to 150 hours. Following milling, the powder is annealed under an inert gas flow at temperatures of 1400°C for six hours. The mechanism of this process is not known, but it is thought that the ball milling process forms nanotube nuclei, and the annealing process activates nanotube growth. Research has shown that this method produces more multi walled nanotubes and few single walled nanotubes.

Other Methods

CNTs can also be produced by diffusion flame synthesis, electrolysis, use of solar energy, heat treatment of a polymer, and low-temperature solid pyrolysis. In flame synthesis, combustion of a portion of the hydrocarbon gas provides the elevated temperature required, with the remaining fuel conveniently serving as the required hydrocarbon reagent. Hence the flame constitutes an efficient source of both energy and hydrocarbon raw material. Combustion synthesis has been shown to be scalable for high-volume commercial production.

Purification Methods

Purification of CNTs generally refers to the separation of CNTs from other entities, such as carbon nanoparticles, amorphous carbon, residual catalyst, and other unwanted species. The classic chemical techniques for purification have been tried, but they have not been found to be effective in removing the undesirable impurities. Three basic methods have been used with varying degrees of success, namely gas-phase, liquid-phase, and intercalation methods.

Generally, a centrifugal separation is necessary to concentrate the single walled nanotubes in low-yield soot before the micro filtration operation, since the nanoparticles easily contaminate membrane filters. The advantage of this method is that unwanted nanoparticles and amorphous carbon are removed simultaneously and the CNTs are not chemically modified. However 2-3 mol nitric acid is useful for chemically removing impurities.

It is now possible to cut CNTs into smaller segments, by extended sonication in concentrated acid mixtures. The resulting CNTs form a colloidal suspension in solvents. They can be deposited on substrates, or further manipulated in solution, and can have many different functional groups attached to the ends and sides of the CNTs.

Gas Phase

The first successful technique for purification of nanotubes was developed by Thomas Ebbesen and coworkers. Following the demonstration that nanotubes could be selectively attached by oxidizing gases these workers realized that nanoparticles, with their defect rich structures might be oxidised more readily than the relatively perfect nanotubes. They found that a significant relative enrichment of nanotubes could be achieved this way, but only at the expense of losing the majority of the original sample.

A new gas-phase method has been developed at the NASA Glenn Research Center to purify gram-scale quantities of single-wall CNTs. This method, a modification of a gas-phase purification technique previously reported by Smalley and others, uses a combination of high-temperature oxidations and repeated extractions with nitric and hydrochloric acid. This improved procedure significantly reduces the amount of impurities such as residual catalyst, and non-nanotube forms of carbon) within the CNTs, increasing their stability significantly.

Liquid Phase

The current liquid-phase purification procedure follows certain essential steps:

preliminary filtration- to get rid of large graphite particles;

dissolution- to remove fullerenes (in organic solvents) and catalyst particles (in concentrated acids)

centrifugal separation-

microfiltration- and

chromatography

It is important to keep the CNTs well-separated in solution, so the CNTs are typically dispersed using a surfactant prior to the last stage of separation.

Intercalation

An alternative approach to purifying multi walled nanotubes was introduced in 1994 by a Japanese research group. This technique made use of the fact that nanoparticles and other graphitic contaminants have relatively "open" structures and can therefore be more readily intercalated with a variety of materials that can close nanotubes.

By intercalating with copper chloride, and then reducing this to metallic copper, the research group was able to preferentially oxidize the nanoparticles away, using copper as an oxidation catalyst. Since 1994, this has become a popular method for purification of nanotubes. "The first stage is to immerse the crude cathodic deposit in a molten copper chloride and potassium chloride mixture at 400oC and leave it for one week. The product of this treatment, which contains intercalated nanoparticles and graphitic fragments, is then washed in ion exchanged water to remove excess copper chloride and potassium chloride.

In order to reduce the intercalated copper chloride-potassium chloride metal, the washed product is slowly heated to 500oC in a mixture of Helium and hydrogen and held at this temperature for 1 hour. Finally, the material is oxidized in flowing air at a rate of 10oC/min to a temperature of 555oC. Samples of cathodic soot which have been treated this way consist almost entirely of nanotubes. A disadvantage of this method is that some amount of nanotubes are inevitably lost in the oxidation stage, and the final material may be contaminated with residues of intercalates. A similar purification technique, which involves intercalation with bromine followed by oxidation, has also been described.

DISPERSION

Although both probe style and bath style ultrasonic systems can be used for dispersing CNTs, it is widely believed that the probe style ultrasonic systems work better for dispersing CNTs. It is also widely known that adding a dispersing reagent (surfactant) into the solution will accelerate the dispersion effect. The reagent Polyvinyl Pyrrolidone (PVP) is a good dispersion agent. Some people like to use the reagent Sodium Dodecyl Benzene Sulfonate (SDBS) or Poly Vinyl Alcohol (PVA) as well. The dispersing reagent and proportions listed above do change when using different solvents. Typically, it is a question of chemistry to achieve a stable dispersion. A stable dispersion will last for days, weeks, or months with little to no settling.

In some applications, achieving a stable dispersion can require other agents in the solution to prevent the CNTs from falling out of solution over time. Emulsifier T-60 (also known as Tween 60) is commonly used with Di water or Isopropyl Alcohol. Organic titanates can be used with Acetone and Xylene. The specific application determines whether these agents remain in the solution when further processing, or if they need to be removed. Some organic titanates can be removed by heating the solution above 250°C. The addition of the OH and COOH functional groups assists the CNTs dispersing in DI water and other solvents as well as the chemical bonding to other materials during further processing.

FUNCTIONALIZATION

Pristine nanotubes are unfortunately insoluble in many liquids such as water, polymer resins, and most solvents. Thus they are difficult to evenly disperse in a liquid matrix such as epoxies and other polymers. This complicates efforts to utilize the nanotubes' outstanding physical properties in the manufacture of composite materials, as well as in other practical applications which require preparation of uniform mixtures of CNTs with many different organic, inorganic, and polymeric materials.

To make nanotubes more easily dispersible in liquids, it is necessary to physically or chemically attach certain molecules, or functional groups, to their smooth sidewalls without significantly changing the nanotubes' desirable properties. This process is called functionalization. The production of robust composite materials requires strong covalent chemical bonding between the filler particles and the polymer matrix, rather than the much weaker van der Waals physical bonds which occur if the CNTs are not properly functionalized.

Functionalization methods such as chopping, oxidation, and "wrapping" of the CNTs in certain polymers can create more active bonding sites on the surface of the nanotubes. For biological uses, CNTs can be functionalized by attaching biological molecules, such as lipids, proteins, biotins, etc. to them. Then they can usefully mimic certain biological functions, such as protein adsorption, and bind to DNA and drug molecules.

This would enable medially and commercially significant applications such as gene therapy and drug delivery. In biochemical and chemical applications such as the development of very specific biosensors, molecules such as carboxylic acid (COOH), poly m-aminobenzoic sulfonic acid (PABS), polyimide, and polyvinyl alcohol (PVA) have been used to functionalize CNTs, as have amino acid derivatives, halogens, and compounds. Some types of functionalized CNTs are soluble in water and other highly polar, aqueous solvents.

PROPERTIES

This section tries to give an overview of the many useful and unique properties of CNTs.

Electrical Conductivity

CNTs can be highly conducting, and hence can be said to be metallic. Their conductivity has been shown to be a function of their chirality, the degree of twist as well as their diameter. CNTs can be either metallic or semi-conducting in their electrical behavior. Conductivity in MWNTs is quite complex. Some types of "armchair"-structured CNTs appear to conduct better than other metallic CNTs. Furthermore, interwall reactions within multi walled nanotubes have been found to redistribute the current over individual tubes non-uniformly. However, there is no change in current across different parts of metallic single-walled nanotubes. The behavior of the ropes of semi-conducting single walled nanotubes is different, in that the transport current changes abruptly at various positions on the CNTs.

The conductivity and resistivity of ropes of single walled nanotubes has been measured by placing electrodes at different parts of the CNTs. The resistivity of the single walled nanotubes ropes was of the order of 10-4 ohm-cm at 27°C. This means that single walled nanotube ropes are the most conductive carbon fibers known. The current density that was possible to achieve was 10-7 A/cm2, however in theory the single walled nanotube ropes should be able to sustain much higher stable current densities, as high as 10-13 A/cm2.

It has been reported that individual single walled nanotubes may contain defects. Fortuitously, these defects allow the single walled nanotubes to act as transistors. Likewise, joining CNTs together may form transistor-like devices. A nanotube with a natural junction (where a straight metallic section is joined to a chiral semiconducting section) behaves as a rectifying diode - that is, a half-transistor in a single molecule. It has also recently been reported that single walled nanotubes can route electrical signals at speeds up to 10 GHz when used as interconnects on semi-conducting devices.

Strength and Elasticity

The carbon atoms of a single sheet of graphite form a planar honeycomb lattice, in which each atom is connected via a strong chemical bond to three neigbouring atoms. Because of these strong bonds, the basal plane elastic modulus of graphite is one of the largest of any known material. For this reason, CNTs are expected to be the ultimate high-strength fibers. Single walled nanotubes are stiffer than steel, and are very resistant to damage from physical forces. Pressing on the tip of a nanotube will cause it to bend, but without damage to the tip. When the force is removed, the nanotube returns to its original state. This property makes CNTs very useful as probe tips for very high-resolution scanning probe microscopy. Quantifying these effects has been rather difficult, and an exact numerical value has not been agreed upon.

Using atomic force microscopy, the unanchored ends of a freestanding nanotube can be pushed out of their equilibrium position, and the force required to push the nanotube can be measured. The current Young's modulus value of single walled nanotubes is about 1 TeraPascal, but this value has been widely disputed, and a value as high as 1.8 Tpa has been reported. Other values significantly higher than that have also been reported.

The differences probably arise through different experimental measurement techniques. Others have shown theoretically that the Young's modulus depends on the size and chirality of the single walled nanotubes, ranging from 1.22 Tpa to 1.26 Tpa. They have calculated a value of 1.09 Tpa for a generic nanotube. However, when working with different multi walled nanotubes, others have noted that the modulus measurements of multi walled nanotubes using AFM techniques do not strongly depend on the diameter. Instead, they argue that the modulus of the multi walled nanotubes correlates to the amount of disorder in the nanotube walls. Not surprisingly, when multi walled nanotubes break, the outermost layers break first.

Thermal Conductivity and Expansion

CNTs have been shown to exhibit superconductivity below 20°K (approx. -253°C). Research suggests that these exotic strands, already heralded for their unparalleled strength and unique ability to adopt the electrical properties of either semiconductors or perfect metals, may someday also find applications as miniature heat conduits in a host of devices and materials.

The strong in-plane graphitic carbon- carbon bonds make them exceptionally strong and stiff against axial strains. The almost zero in-plane thermal expansion but large inter-plane expansion of single walled nanotubes implies strong in-plane coupling and high flexibility against non-axial strains. Many applications of CNTs, such as in nanoscale molecular electronics, sensing and actuating devices, or as reinforcing additive fibers in functional composite materials, have been proposed. Reports of several recent experiments on the preparation and mechanical characterization of CNT-polymer composites have also appeared. These measurements suggest modest enhancements in strength characteristics of CNT-embedded matrixes as compared to bare polymer matrixes. Preliminary experiments and simulation studies on the thermal properties of CNTs show very high thermal conductivity. It is expected, therefore, that nanotube reinforcements in polymeric materials may also significantly improve the thermal and thermomechanical properties of the composites.

Field Emission

Field emission results from the tunneling of electrons from a metal tip into vacuum, under application of a strong electric field. The small diameter and high aspect ratio of CNTs is very favourable for field emission. Even for moderate voltages, a strong electric field develops at the free end of supported CNTs because of their sharpness.

This was observed by de Heer and co-workers at EPFL in 1995. He also immediately realized that these field emitters must be superior to conventional electron sources and might find their way into all kind of applications, most importantly flat-panel displays. It is remarkable that after only five years Samsung actually realized a very bright colour display, which will be shortly commercialized using this technology. Studying the field emission properties of multi walled nanotubes, Bonard and co-workers at EPFL observed that together with electrons, light is emitted as well. This luminescence is induced by the electron field emission, since it is not detected without applied potential. This light emission occurs in the visible part of the spectrum, and can sometimes be seen with the naked eye.

High Aspect Ratio

CNTs represent a very small, high aspect ratio conductive additive for plastics of all types. Their high aspect ratio means that a lower loading of CNTs is needed compared to other conductive additives to achieve the same electrical conductivity. This low loading preserves more of the polymer resins' toughness, especially at low temperatures, as well as maintaining other key performance properties of the matrix resin. CNTs have proven to be an excellent additive to impart electrical conductivity in plastics. Their high aspect ratio, about 1000:1 imparts electrical conductivity at lower loadings, compared to conventional additive materials such as carbon black, chopped carbon fiber, or stainless steel fiber.

Highly Absorbent

The large surface area and high absorbency of CNTs make them ideal candidates for use in air, gas, and water filtration. A lot of research is being done in replacing activated charcoal with CNTs in certain ultra high purity applications.

THE APPLICATIONS

The special nature of carbon combined with the molecular perfection of single-walled nanotubes to endow them with exceptional material properties, such as very high electrical and thermal conductivity, strength, stiffness, and toughness. No other element in the periodic table bonds to itself in an extended network with the strength of the carbon-carbon bond. The delocalized pi-electron donated by each atom is free to move about the entire structure, rather than remain with its donor atom, giving rise to the first known molecule with metallic-type electrical conductivity.

Furthermore, the high-frequency carbon-carbon bonds vibrations provide an intrinsic thermal conductivity higher than even diamond. In most conventional materials, however, the actual observed material properties-strength, electrical conductivity, etc.- are degraded very substantially by the occurrence of defects in their structure. For example, high-strength steel typically fails at only about 1% of its theoretical breaking strength. CNTs, however, achieve values very close to their theoretical limits because of their molecular perfection of structure.

This aspect is part of the unique story of CNTs. CNTs are an example of true nanotechnology: they are under 100 nanometers in diameter, but are molecules that can be manipulated chemically and physically in very useful ways. They open an incredible range of applications in materials science, electronics, chemical processing, energy management, and many other fields. CNTs have extraordinary electrical conductivity, heat conductivity, and mechanical properties. They are probably the best electron field-emitter possible.

They are polymers of pure carbon and can be reacted and manipulated using the well-known and the tremendously rich chemistry of carbon. This provides opportunity to modify their structure, and to optimize their solubility and dispersion. Very significantly, CNTs are molecularly perfect, which means that they are normally free of property-degrading flaws in the nanotube structure. Their material properties can therefore approach closely the very high levels intrinsic to them. These extraordinary characteristics give CNTs potential in numerous applications.

Field Emission

CNTs are the best known field emitters of any material. This is understandable, given their high electrical conductivity, and the incredible sharpness of their tip. The smaller the tip's radius of curvature, the more concentrated the electric field will be, leading to increased field emission. The sharpness of the tip also means that they emit at especially low voltage, an important fact for building low-power electrical devices that utilize this feature. CNTs can carry an astonishingly high current density. Furthermore, the current is extremely stable.

An immediate application of this behavior receiving considerable interest is in field-emission flat-panel displays. Instead of a single electron gun, as in a traditional cathode ray tube display, in CNT-based displays there is a separate nanotube electron gun for each individual pixel in the display. Their high current density, low turn-on and operating voltages, and steady, long-lived behavior make CNTs very attractive field emitters in this application. Other applications utilizing the field-emission characteristics of CNTs include general types of low-voltage cold-cathode lighting sources, lightning arrestors, and electron microscope sources.

Conductive or Reinforced Plastics

Much of the history of plastics over the last half-century has involved their use as a replacement for metals. For structural applications, plastics have made tremendous headway, but not where electrical conductivity is required, because plastics are very good electrical insulators. This deficiency is overcome by loading plastics up with conductive fillers, such as carbon black and larger graphite fibers. The loading required to provide the necessary conductivity using conventional fillers is typically high, however, resulting in heavy parts, and more importantly, plastic parts whose structural properties are highly degraded. It is well-established that the higher the aspect ratio of the filler particles, the lower the loading required to achieve a given level of conductivity. CNTs are ideal in this sense, since they have the highest aspect ratio of any carbon fiber. In addition, their natural tendency to form ropes provides inherently very long conductive pathways even at ultra-low loadings. Applications that exploit this behavior of CNTs include EMI/RFI shielding composites; coatings for enclosures, gaskets, and other uses such as electrostatic dissipation; antistatic materials, transparent conductive coatings; and radar-absorbing materials for stealth applications.

A lot of automotive plastics companies are using CNTs as well. CNTs have been added into the side mirror plastics on automobiles in the US since the late 1990s. I have seen forecasts predicting that GM alone will consume over 500 pounds of CNT masterbatches in 2006 for using in all areas of automotive plastics. Masterbatches normally contain 20 wt% cnts which are already very well dispersed. Manufacturers then need to perform a "let down" or dilution procedure prior to using the masterbatch in production

Energy Storage

CNTs have the intrinsic characteristics desired in material used as electrodes in batteries and capacitors, two technologies of rapidly increasing importance. CNTs have a tremendously high surface area, good electrical conductivity, and very importantly, their linear geometry makes their surface highly accessible to the electrolyte. Research has shown that CNTs have the highest reversible capacity of any carbon material for use in lithium ion batteries. In addition, CNTs are outstanding materials for super capacitor electrodes and are now being marketed for this application. CNTs also have applications in a variety of fuel cell components. They have a number of properties, including high surface area and thermal conductivity, which make them useful as electrode catalyst supports in PEM fuel cells. Because of their high electrical conductivity, they may also be used in gas diffusion layers, as well as current collectors. CNTs' high strength and toughness-to-weight characteristics may also prove valuable as part of composite components in fuel cells that are deployed in transport applications, where durability is extremely important.

Conductive Adhesives and Connectors

The same properties that make CNTs attractive as conductive fillers for use in electromagnetic shielding, ESD materials, etc., make them attractive for electronics packaging and interconnection applications, such as adhesives, potting compounds, coaxial cables, and other types of connectors.

Molecular Electronics

The idea of building electronic circuits out of the essential building blocks of materials- molecules- has seen a revival the past few years, and is a key component of nanotechnology. In any electronic circuit, but particularly as dimensions shrink to the nanoscale, the interconnections between switches and other active devices become increasingly important. Their geometry, electrical conductivity, and ability to be precisely derived, make CNTs the ideal candidates for the connections in molecular electronics. In addition, they have been demonstrated as switches themselves. There are already companies such as Nantero from Woburn, MA that are already making CNT based non-volitle random access memory for PC's. A lot of research is being done to design CNT based transistors as well.

Thermal Materials

The record-setting anisotropic thermal conductivity of CNTs is enabling many applications where heat needs to move from one place to another. Such an application is found in electronics, particularly heat sinks for chips used in advanced computing, where uncooled chips now routinely reach over 100oC. The technology for creating aligned structures and ribbons of CNTs [D.Walters, et al., Chem. Phys. Lett. 338, 14 (2001)] is a step toward realizing incredibly efficient heat conduits. In addition, composites with CNTs have been shown to dramatically increase their bulk thermal conductivity, even at very small loadings.

Structural Composites

The superior properties of CNTs are not limited to electrical and thermal conductivities, but also include mechanical properties, such as stiffness, toughness, and strength. These properties lead to a wealth of applications exploiting them, including advanced composites requiring high values of one or more of these properties.

Fibers and Fabrics

Fibers spun of pure CNTs have recently been demonstrated and are undergoing rapid development, along with CNT composite fibers. Such super-strong fibers will have many applications including body and vehicle armor, transmission line cables, woven fabrics and textiles.

Catalyst Support

CNTs intrinsically have an enormously high surface area; in fact, for single walled nanotubes every atom is not just on one surface- each atom is on two surfaces, the inside and the outside of the nanotube! Combined with the ability to attach essentially any chemical species to their sidewalls this provides an opportunity for unique catalyst supports. Their electrical conductivity may also be exploited in the search for new catalysts and catalytic behavior.

CNT Ceramics

A ceramic material reinforced with carbon nanotubes has been made by materials scientists at UC Davis. The new material is far tougher than conventional ceramics, conducts electricity and can both conduct heat and act as a thermal barrier, depending on the orientation of the nanotubes. Ceramic materials are very hard and resistant to heat and chemical attack, making them useful for applications such as coating turbine blades, but they are also very brittle. The researchers mixed powdered alumina (aluminum oxide) with 5 to 10 percent carbon nanotubes and a further 5 percent finely milled niobium. The researchers treated the mixture with an electrical pulse in a process called spark-plasma sintering. This process consolidates ceramic powders more quickly and at lower temperatures than conventional processes.

The new material has up to five times the fracture toughness - resistance to cracking under stress - of conventional alumina. The material shows electrical conductivity seven times that of previous ceramics made with nanotubes. It also has interesting thermal properties, conducting heat in one direction, along the alignment of the nanotubes, but reflecting heat at right angles to the nanotubes, making it an attractive material for thermal barrier coatings.

Biomedical Applications

The exploration of CNTs in biomedical applications is just underway, but has significant potential. Since a large part of the human body consists of carbon, it is generally though of as a very biocompatible material. Cells have been shown to grow on CNTs, so they appear to have no toxic effect. The cells also do not adhere to the CNTs, potentially giving rise to applications such as coatings for prosthetics and surgical implants. The ability to functionalize the sidewalls of CNTs also leads to biomedical applications such as vascular stents, and neuron growth and regeneration. It has also been shown that a single strand of DNA can be bonded to a nanotube, which can then be successfully inserted into a cell; this has potential applications in gene therapy.

Air, Water and Gas Filtration

Many researchers and corporations have already developed CNT based air and water filtration devices. It has been reported that these filters can not only block the smallest particles but also kill most bacteria. This is another area where CNTs have already been commercialized and products are on the market now. Someday CNTs may be used to filter other liquids such as fuels and lubricants as well. A lot of research is being done in the development of CNT based air and gas filtration. Filtration has been shown to be another area where it is cost effective to use CNTs already. The research I've seen suggests that 1 gram of MWNTs can be dispersed onto 1 sq ft of filter media. Manufacturers can get their cost down to 35 cents per gram of purified MWNTs when purchasing ton quantities.

Other Applications

Some commercial products on the market today utilizing CNTs include stain resistant textiles, CNT reinforced tennis rackets and baseball bats. Companies like Kraft foods are heavily funding cnt based plastic packaging. Food will stay fresh longer if the packaging is less permeable to atmosphere. Coors Brewing company has developed new plastic beer bottles that stay cold for longer periods of time. Samsung already has CNT based flat panel displays on the market. A lot of companies are looking forward to being able to produce transparent conductive coatings and phase out ITO coatings. Samsung uses align SWNTs in the transparent conductive layer of their display manufacturing process.

3

BIG WORLD OF SMALL THINGS

Think small, Really small ! Got your old high school microscope? Not good enough. Your handy computer microchip? Still way too big. Think about building machines only a few molecules or atoms in size-that is one aspect of nanotechnology, a science of the future that is being created today.

Nanotechnology is a term that encompasses scientific and engineering activities at the nanometer scale. A nanometer is one billionth of a meter, or only a few atoms long. Using an array of ultra-precise tools, scientists can create electric components and machines that are virtually invisible to the naked eye. Even genetic engineering and bioremediation, the use of microbes to eliminate hazardous wastes, fall under the broad definition of nanotechnology.

This is not science fiction but science fact. Two years have passed since IBM scientists spelled out their company logo by moving 35 xenon atoms with a device called a scanning- tunneling microscope. Yet in that short time, nanotechnology has demonstrated the potential to become an important, pervasive technology. Today, for example, microsensors tinier than the width of a human hair are routinely being used in automobile anti-lock braking systems.

COMPUTERS IN FUTURE

Few industries can boast the huge leaps in technology that the computer industry has made in the last 50 years. Since the invention of the transistor in the 1940s, computers have shrunk from behemoth machines that took up multiple rooms to portable devices the size of paperback books that can perform hundreds of millions of operations per second. While computer manufacturing has made great strides in the last half-century, the manufacturing process is still limited to a handful of companies.

Manufacturing computers is a costly and time-consuming undertaking. A microprocessor fabrication plant costs around $2 billion and takes two full weeks to produce one silicon-based microprocessor. Few computer enthusiasts have the resources to make their own computer chips. However, researchers are developing ways to allow anyone to become their own microprocessor fabricator. Users will simply download microchip designs from the Internet and print out a working ink-based, plastic processor on a desktop fabrication machine, similar to an ink jet printer. The next phase of computing will make the users into the creators and builders of their own computer components.

THE DESKTOP FABRICATION

Few argue that the next generation of computers will be nearly invisible, meaning that they will blend in with everyday objects. Flexible ink-like circuitry will be printed onto plastic or sprayed onto various other substrates, such as clothes. One of the scientists leading this printable computer revolution is Joseph Jacobson of MIT Media Lab's Nano Media Group. Jacobson has said that his group will be able to produce a simple printed microprocessor in late 2001 or early 2002. He also foresees being able to eventually produce a printed chip that could rival an Intel Pentium processor.

Jacobson's group has already succeeded in using an ordinary Hitachi ink jet printer to make several components for a printable computer. Using a nanoparticle-based ink made from suspending nano-size semiconductor particles in a liquid, researchers spray the components onto a plastic substrate. Here's a look at some of the printed components the MIT group has made with this process:

- *Thermal actuators* - An actuator is a sensor that causes a device to be turned on, off, adjusted or moved. In a thermal actuator, heat is used to cause the expansion of components to create movement.
- *Linear-drive motors* - This type of motor is similar to a normal electric motor, which has a magnet that circles around the coil loop to make the motor spin. However, there is one key difference. Think of linear-drive motors as flattened electric motors containing a flat magnet moving back and forth across a coil. In a sense, the magnet in the linear drive motor acts like a piston.
- *Microelectromechanical Systems (MEMS)* - MEMS are touted as the precursor or bridge to nanotechnology. These micromachines are used in a variety of devices, including pacemakers, games, and accelerometers of airbags. They perform a variety of functions, including sensing, communication and actuation. In the future, MEMS are expected to have the ability to self-replicate.

The Media Lab also created transistors using a different process. For that, polymer stamps are used with the architecture of the transistors in a positive relief. The stamp is then dipped in the nanoparticle ink and transferred to a substrate by hand. The next step will be to use an ink jet printer or some other kind of desktop fabricator to create printable transistors. MIT isn't the only group developing ways to print computer circuitry. Plastic Logic, a company that sprang out of work begun at Cambridge University in England, plans to market the first plastic chip. The company has developed and patented a method of printing plastic onto polymer substrates, making cheap and flexible plastic transistors. The process is similar to the ink jet process used by MIT, but Plastic Logic adds carbon-based chemicals to alter the properties of the plastic. By printing the chips onto rolls of film, they can be applied to a variety of surfaces.

At Lucent Technologies' research company, Bell Labs Innovations, researchers developed the world's first printed transistor in 1997. Using plastic sheets similar to overhead projector transparencies, a liquid plastic semiconductor is applied over a stainless-steel mesh with a squeegee to form the multiple layers of the transistor. After the solvent of the mixture evaporates, the plastic remains. The process is very similar to how silk screening works. Lucent has teamed up with E Ink, an MIT offspring, to create printable displays. See How Electronic Ink Will Work for more information. Soon, scientists will be able to create just about every part of a computer's hardware using a desktop fabricator. Plastic will take the place of silicon for many purposes, but don't expect to write off silicon as a valuable computer component for at least a decade or two. In the next section, we'll see how plastic stacks up against silicon and why we can expect silicon to stick around for many more years.

PLASTIC VS. SILICON

Plastic may revolutionize the semiconductor industry, but it won't be an overnight revolution. The sophistication of printable computers is still very simple. Currently, plastic fabrication devices can only produce transistors at the 25 micrometer scale (a micrometer is one-millionth of a meter); that's far from the .2 micrometer resolution that is needed to create a working microprocessor. Intel is able to crowd about 10 million transistors only a few hundred nanometers big onto one silicon chip. A nanometer is one-billionth of a meter.

Most researchers will tell you that printable computer components are not designed to replace silicon. Initially, we will see these printable devices used to give intelligence to everyday objects. They will be integrated into clothes, food labels and toys. One of the most exciting applications for printable electronics is creating a wallpaper that doubles as a television screen or computer monitor. MIT also plans to build a digital camera into a a business card.

Plastic does offer some benefits over silicon. Silicon is rigid, while plastic chips are flexible, allowing it to be placed on a variety of substrates. The problem is that, despite great hopes to create a plastic Pentium, printed inorganic transistors are still about 100 times slower than conventional transistors found on silicon chips.

Basically, printable computers represent the merging of conventional printing technologies with computer chip fabrication to produce cheaper, more flexible components. While many obstacles remain in its development, early products are ready to enter the market, such as disposable cell phones and computerized clothing. The next decade may bring us the ability to print out our own electronic devices and sophisticated computers.

NANOTECH GONE BAD

While one would hope that the slightly eccentric and amoral mad scientist is a thing of fiction, the possibility for scientists to get it wrong, or to lose control of their new creations, remains grounded in reality. History has shown us that the development of new materials and technologies has led to the disruption and destruction of countless lives. But what can we do? Nobody has a crystal ball. All technology has a cost/benefit ratio, and the trick has been to predict what the likely risks will be. Who could have predicted the effects that the automobile industry would have on the environment? Who can confidently say what the fallout from current technologies like cell phones will be? We just have to accept that while we mostly enjoy the lives that science brings us, any exciting new technology could open up a Pandora's Box of horrors. But if we were to play psychic, which advancing field of science is likely to be the next asbestos or thalidomide? The smart money, without a doubt, would be on nanotechnology.

Nanotechnology is poised to be humanity's greatest technological achievement, or its greatest blunder. Once deemed some kind of grant garnering scheme unlikely to ever bear fruit, nanotechnology's profile has risen sharply of late. In fact, nanotech's preeminent rise is marked by the steady exodus of scientists from other fields of science into nanoscience. Scientists Andreas Barth, from the FIZ Karlsruhe, and Werner Marx, from the Max Planck Institute for Solid-State Research, recently stated that unless a groundbreaking discovery was made soon, high-temperature superconductivity would be a dead field within four years. The researchers laid the blame for the brain-drain on nanoscience, citing as an example the avalanche of papers relating to nanotubes compared to those about high-temperature superconductors.

One of those at the forefront of nanoscience is Keith Schwab, who earlier this year moved from the National Security Agency to take up the position of associate professor of physics at Cornell University. Schwab's area of research tests the very boundary (if there is one) between the quantum domain and the world of Newton's classical mechanics. Schwab has developed a nanomechanical device so small that it appears to be on the very edge of the two worlds, otherwise known as the quantum mechanical limit. The device is comprised of a sliver of aluminum on silicon nitride, and is 8.7 microns (millionths of a meter) long and 200 nanometers (billionths of a meter) wide.

It sounds small, but it's many times bigger than your garden variety quantum object. But Schwab's device still manages to push the limits set by Heisenberg's uncertainty principle. The principle states that the accuracy of simultaneous measurements of position and velocity of a particle is limited by a quantifiable amount. In this case, Schwab was able to get closer to this theoretical limit than anyone previously. The weird nature of quantum effects was made manifest as the team's observation triggered movement in the apparatus (a phenomenon known as "back action").

"We made measurements of position that are so intense- so strongly coupled- that by looking at it we can make it move," said Schwab. "Quantum mechanics requires that you cannot make a measurement of something and not perturb it. We're doing measurements that are very close to the uncertainty principle; and we can couple so strongly that by measuring the position we can see the thing move."

Spooky stuff. And Schwab wants to push things further, by building a nanomechanical device that exhibits superposition principle properties. In other words, a nano-device that can be in two places at once. "What's really neat is it looks like we should be able to do it," Schwab said. "The hope, the dream, the fantasy, is that we get that superposition and start making bigger devices and find the breakdown." Schwab's continued research on the mysteries that lie between the quantum and classical worlds is tipped to deliver significant advances in quantum computing, cooling engineering, communications and medicine. The list of applications could be unlimited.

There's no doubt that the development of micro-machines the size of a molecule will affect humanity as fundamentally as the Industrial Revolution did. But like any other technology, there are going to be some downsides, says The Nanoethics Group, an industry research and education organization. "If we had given foresight to how the invention or discovery of electricity, factories, automobiles, nuclear power and the Internet might affect people and society, we might have done a much better job in managing their negative consequences-such as economic disruption, urban sprawl, pollution, nuclear arms race and high-tech crimes," explained Patrick Lin, research director for The Nanoethics Group.

Groups such as Nanoethics are not saying that nanotechnology is something to be shunned, but they are saying that we should be on the defensive in regard to the possible use or misuse of nanotechnologies. While breakthroughs in nanotech might seem "really neat" to some, Nanoethics wants to actually think through the implications of such advances.

Importantly, concerns about nanotechnology have moved on from the fanciful bleatings of sci-fi obsessed doomsayers. Sure, the military, privacy and God syndrome anxieties are still relevant, but some of the real concerns expressed by scientists are a little less... well, exciting. Specific concerns about nanotechnology have come from no less than the UK Royal Society and Royal Academy of Engineering, who suggest that exposure to manufactured nano-sized particles may be detrimental to human health. Andrew Maynard, from the Woodrow Wilson International Center for Scholars, expresses similar concerns. Coupled with environmental worries, Maynard believes that nano particles could have an unprecedented reactivity to the human body, as their size, structure and purpose may easily evade the body's natural defense systems. "We need to understand both how harmful a substance is, and how much of it can get into the body, if risk is to be understood and managed," says Maynard.

Jared Diamond says that we can't always rely on technology to get us out of a tight spot. Diamond's argument summons up an image of humanity as a character using its fingers and toes in a desperate attempt to plug up the rapidly appearing holes of a bursting dam. Humanity has so far had enough appendages to plug up the holes created by our technologies, but what new technology will be available to save us if our nano-creations go haywire?

The Real-World Quantum Effects

The realm of quantum mechanics met the world of classical physics when an antenna-like sliver of silicon one-tenth the width of a human hair oscillated in a Boston University lab. With two sets of protrusions, much like the teeth from a two-sided comb, the antenna not only exhibits the first quantum nanomechanical motion but is also the world's fastest moving nanostructure. Comprised of 50 billion atoms, the antenna built is so far the largest structure to display quantum mechanical movements. The device is also the fastest of its kind, oscillating at 1.49 gigahertz, or 1.49 billion times a second. The quantum effect is evident when the nanomechanical oscillator starts to jump between two discrete positions without occupying the physical space in between.

Physicists led by Pritiraj Mohanty developed the nanomechanical oscillator at Boston University. Operating at gigahertz speeds, it could help further miniaturize wireless communication devices, but more important to the researchers, the oscillator lies at the cusp of traditional physics and quantum physics. "It's a truly macroscopic quantum system," says Alexei Gaidarzhy, the paper's lead author.

Because the nanomechanical oscillator is "large", the research team was able to attach electrical wiring to its surface in order to monitor tiny discrete quantum motion, behavior that exists in the realm of atoms and molecules. At a certain frequency, the paddles begin to vibrate in concert, causing the central beam to move at that same high frequency, but at an increased and easily measured amplitude. Where each paddle moves only about a femtometer, roughly the diameter of an atom's nucleus, the antenna moves over a distance of one-tenth of a picometer, a tiny distance that still translates to a 100-fold increase in amplitude.

When fabricating and testing the nanomechanical device, the researchers placed the entire apparatus, including the cryostat and monitoring devices, in a copper-walled, copper-floored room. This set-up shielded the experiment from unwanted vibration noise and electromagnetic radiation that could generate from outside electrical devices, such as cell phone signals, or even the movement of subway trains outside the building. Even with these precautions, performing such novel experiments is tricky. "When it's a new phenomenon, it's best not to be guided by expectations based on conventional wisdom," says Gaidarzhy. "The philosophy here is to let the data speak for itself." The group carries out the experiments under extremely cold conditions, at a temperature of 110 millikelvin, which is only a tenth of a degree above the absolute zero. When cooled to such a low temperature, the nanomechanical oscillator starts to jump between two discrete positions without occupying the physical space in between, a telltale sign of quantum behavior.

BIG CONUNDRUMS GROW FROM LITTLE THINGS

Whether you fear it, welcome it, don't understand it or think it an elaborate scientific conspiracy, "molecular nanotechnology" (MNT), or "molecular manufacturing," will have a profound and instantaneous effect on all aspects of industry and society. According to the Center for Responsible Nanotechnology (CRN), the new technology will hit us with all the destructive force of a speeding freight train, and "without adequate preparation and study, the effects could be dangerously disruptive." Yet more science hype? Perhaps. But in 11 essays, published by Nanotechnology Perceptions, a peer-reviewed academic journal of the Collegium Basilea in Basel, Switzerland, a number of highly regarded scientists cite past mistakes involving the introduction of a new technology without adequate public consultation and discussion. They also imagine a number of future nanotech scenarios that both fascinate and frighten.

In one of the essays, Singularities and Nightmares, David Brin offers some positive words on the potential benefits of nanotechnology, such as genome mapping while-you-wait. Perhaps, he says, this could be achieved conveniently by: "compressing a complete biochemical laboratory the size of a house down to something that fits cheaply on your desktop. A MolecuMac, if you will." But on the flip side, Brin makes your hair stand on end when he discusses the possibility of MolecuMac "hackers", whose ranks may include terrorists, despots or even angry, misunderstood, spotty-faced youths hell-bent on making your life a misery. "What are we going to do when kids all over the world can analyze and synthesize any organic compound, at will? What happens when this kind of [hacking] 'creativity' moves to the very stuff of life itself?" asks Brin.

Brin draws attention to influential people in society, such as Francis Fukuyama, author of The End of History, whose opinions are polar-opposite when it comes down to the issue of how much control and regulation there should be on technological innovation. Renunciation is the key word for this group, who believe that in order to save humanity from itself we must show restraint and curtail our creative juices for the sake of a better world. Brin says that authors such as Margaret Attwood and Michael Crichton often express: "outright loathing for the overweening arrogance of hubristic technological innovators who just cannot leave nature well enough alone." And what about former Sun Computers exec Bill Joy, who claims that the notion of an open society is dead? Brin says that Joy uses the Unabomber as reason enough for believing that: "our sole hope for survival may be to renounce, squelch, or relinquish several classes of technological progress." Phew. Brin is supportive of the social critics who shout when they see potential danger along the road, but he doesn't take the argument for containment as far as they do.

Most of the CRN fraternity seems to support the happy medium Brin refers to as "reciprocal accountability"; which advocates openness and accountability for technological development. Firebrand futurologist Ray Kurzweil sums it up when he says: "As the pace of technological advancement rapidly accelerates, it becomes increasingly important to promote knowledgeable and insightful discussion of both promise and peril." This is probably true of any new technology, but especially true in regard to nanotechnology, claims Mike Treder, the Executive Director of the CRN. Which is why, in August 2005, CRN formed a special task force comprised of experts from multiple disciplines whose job it is to: "develop comprehensive recommendations for the safe and responsible use of nanotechnology". But is a collaborative network of big-brained individuals really enough to affect how the future of molecular manufacturing will pan out? The fact is, most other attempts at creating an effective policy to govern new technology have failed or never seen the light of day, because nobody can really predict what applications will stem from new scientific discoveries. And it is this freedom to create without prophetic insight that gets us all into so much trouble. So it stands to reason that there will be some who think humanity is not yet responsible enough to have that freedom. If history is anything to go by, perhaps the neo-luddites are correct and we must make technological sacrifices to save ourselves, as the halfway policy of reciprocal accountability just won't cut it.

It would be hard to imagine, for example, President Truman saying to the A-bomb team, "ok, boys, wrap it up, I'm losing sleep over this technology." And what about all those private institutes, whose doors are more often than not closed to public or government scrutiny? Just what will be in your face cream or toothpaste in the years to come? It's hard to imagine just how a nanotech policy would work in advance of any future applications. And who's to say that in trying to prevent the development of an undesirable application, a policy panel wouldn't also stifle positive innovation and progress? The industrial age seems to have a lot to answer for in regard to climate change these days, but would any of us have prevented the development of the internal combustion engine? Should we have turned our backs on electricity, because Harold P. Brown invented the electric chair? And would we have sacrificed radioactivity at the expense of some of the more advanced medical technologies? It's an unfortunate fact that very often the applications stemming from scientific discoveries are Janus faced. And unfortunately, no god has endowed anyone from CRN with second sight.

If nanotechnology proves to live up to all the hype, and personal nanofactories become the norm, it will be the most versatile and pervasive technology ever conceived by humanity. As such, it is unlikely that any meaningful guidelines could be imposed upon research and development for any future molecular manufacturing applications. While the CRN's Task Force is a welcome initiative, regulating potential applications once molecular manufacturing is in full swing will become analogous to bailing out a sinking boat with a sieve.

RNA Used as Nano-Scaffolding

Microscopic scaffolding to house the tiny components of nanotech devices could be built from ribonucleic acid (RNA), the same substance that transports messages around a cell's nucleus, says a research group at Purdue University. By encouraging RNA molecules to self-assemble into 3-D shapes resembling spirals, triangles, rods and hairpins, the group has found what could be a method of constructing lattices on which to build complex microscopic machines. From such RNA blocks, the group has already constructed arrays that are several micrometers in diameter.

"Our work shows that we can control the construction of three-dimensional arrays made from RNA blocks of different shapes and sizes," said Peixuan Guo, from Purdue's School of Veterinary Medicine. "With further research, RNA could form the superstructures for tomorrow's nanomachines." The research appears in the journal Nano Letters. Nanotechnologists, like those in Guo's group, hope to build microscopic devices with sizes that are best measured in nanometers- or billionths of a meter. Because nature routinely creates nano-sized structures for living things, many researchers are turning to biology for their inspiration and construction tools. "Biology builds beautiful nanoscale structures, and we'd like to borrow some of them for nanotechnology," Guo said. "The trouble is, when we're working with such tiny blocks, we are short of tiny steam shovels to push them around. So we need to design and construct materials that can assemble themselves."

Organisms are built in large part of three main types of building blocks: proteins, DNA and RNA. Of the three, perhaps least investigated and understood is RNA, a molecular cousin to the DNA that stores genetic blueprints within our cells' nuclei. RNA typically receives less attention than other substances from many nanotechnologists, but Guo said the molecule has distinct advantages. "RNA combines the advantages of both DNA and proteins and puts them at the nanotechnologist's disposal," Guo said. "It forms versatile structures that are also easy to produce, manipulate and engineer."

Since his discovery of a novel RNA that plays a vital role in a microscopic "motor" used by the bacterial virus phi29, Guo has continued to study the structure of this RNA molecule for years. It formed the "pistons" of a tiny motor his lab created several years ago, and members of the team collaborated previously to build dimers and trimers- molecules formed from two and three RNA strands, respectively. "By designing sets of matching RNA molecules, we can program RNA building blocks to bind to each other in precisely defined ways," he said. "We can get them to form the nano-shapes we want."

From the small shapes that RNA can form- hoops, triangles and so forth- larger, more elaborate structures can in turn be constructed, such as rods gathered into spindly, many-pronged bundles. These structures could theoretically form the scaffolding on which other components, such as nano-sized transistors, wires or sensors, could be mounted.

"Because these RNA structures can be engineered to put themselves together, they could be useful to industrial and medical specialists, who will appreciate their ease of engineering and handling," said co-researcher Dieter Moll. "Self-assembly means cost-effective."

Moll, while bullish on RNA's prospects, cautioned that there was more work to be done before nanoscale models could be built at will. "One of our main concerns right now is that, over time, RNA tends to degrade biologically," he said. "We are already working on ways to make it more resistant to degradation so that it can form long-lasting structures."

"We have not built actual scaffolds yet, just 3-D arrays," he said. "But we have built them from engineered biological molecules, and that could help us bridge the gap between the living and the nonliving world. If nanotech devices can eventually be built from both organic and inorganic materials, it would ease their use in both medical and industrial settings, which could multiply their usefulness considerably."

4

NANOTECHNOLOGY AND CHIP TECHNOLOGY

For years the public has viewed Nanotechnology as the stuff of Science Fiction - small robots a millionth of an inch in size, capable of transforming the very molecular structure of matter. Whether it's Neal Stephenson's SF novel "The Diamond Age" describing nano-probes building entire cities from just the raw materials, or Start Trek's Borg transforming people into cyborg drones, nanotechnology has appeared to the public as "way out there." However, nanotech has a reality in scientific circles, and a number of companies are starting to reap financial rewards from being early innovators.

Physicist Richard Feynman first suggested the concept of nanotechnology in 1959. In a famous speech he first suggested that devices and materials could be assembled atom-by atom. The field has been nurtured and evangelized by K. Eric Drexler of the Foresight Institute, ever since his 1981 journal article on molecular nanotechnology. The ability to build things on a molecular level- down to nanometer size - will require a whole new set of technologies, as dealing with the ultra-small involves a different set of rules. However, once an "Assembler" which can build things atom-by-atom has become a reality, almost any device that could be conceived (designed and programmed) could be built cheaply from raw materials. Nanometer devices are not as far away as you might think. Scientists and engineers have learned to create micron-scale "micro-tech" devices of about a thousandth of an inch in size. These MicroElectroMechanical Systems (MEMS) can be built with very small moving parts, and is real, commercial technology today. Techniques for working at a micron level with silicon semiconductors can create ultra-fast computing components, ultra-small sensors and telemetry devices, and ultra-high bandwidth communications networks. They will transform the Internet, revolutionize medicine, and bring visions of the future into reality.

We will have three pioneering companies present their visions of how these technologies will reshape the 21st century, and how they plan to profit from them.

Texas Instruments is a pioneer in the MEMS field with their DLP and DMD (Digital Light Processing and Digital Micromirror Device) technology. Integrating a million controllable mirrors or more on a semiconductor chip, each device is capable of directly converting digital electronic information into dramatic visual images. Today the transport, as well as the display, of high bandwidth information is increasingly done in the photonic domain. MEMS technology has broad applicability to the delivery of high bandwidth content.

Ball Semiconductor has not only invented the techniques for producing semiconductors on small three-dimensional surfaces, they have incorporated MEMS and assembler-like technologies into their plant. On one end of the fabrication line sand is turned into millimeter-sized silicon spheres. Travelling from process to process in plastic tubing, out the other end of the fab will come powerful semiconductor devices.

Zyvex is betting on the future of nanotechnology. Doing basic research into nanotech assembly techniques, their stated goal is to develop the first true Assembler. It may be ten years before their first commercial success, but they have already emerged as the world's first (and perhaps only) commercial nanotechnology company.

HISTORICAL PERSPECTIVE

- In a talk given in 1959, Richard Feynman was the first scientist to suggest that devices and materials could someday be fabricated to atomic specifications: "The principles of physics, as far as I can see, do not speak against the possibility of maneuvering things atom by atom."
- The first journal article published on molecular nanotechnology: "Molecular engineering: An approach to the development of general capabilities for molecular manipulation," Proceedings of the National Academy of Sciences, September 1981, is now available at the Institute for Molecular Manufacturing Web site.
- A short history of the idea of nanotechnology is given in "Nanotechnology: Evolution of the Concept," Journal of the British Interplanetary Society, Vol. 45, pp. 395-400. This essay was reprinted in the book Prospects in Nanotechnology: Toward Molecular Manufacturing, ed. (Markus Krummenacker and James Lewis, Wiley, 1995).

BENEFICIAL NANOTECHNOLOGY

Foresight is the leading think tank and public interest institute on nanotechnology. Founded in 1986, Foresight was the first organization to educate society about the benefits and risks of nanotechnology. At that time, nanotechnology was a little-known concept.

Today, with the basic framework of public understanding in place, we are refocusing our efforts on guiding nanotechnology research, public policy and education to address the critical challenges facing humanity. Foresight's new mission is to ensure the beneficial implementation of nanotechnology.

Foresight is accomplishing this by providing balanced, accurate and timely information to help society understand and utilize nanotechnology through public policy activities, publications, guidelines, networking events, tutorials, conferences, roadmaps and prizes.

Foresight is a member-supported organization. Our membership, including over 14,000 individuals and a growing number of corporations, is diverse demographically and geographically. They are interested in ensuring that the future of nanotechnology unfolds for the benefit of all. These concerned individuals include scientists, engineers, business people, investors, publishers, artists, ethicists, policy makers, interested laypersons, and students from grammar school to graduate level.

Sci-fi in the Packaging Industry

Nanotechnology is the newest craze among media, academia, investors (just think of the ROI if you hit a company that comes up wit the best "micromachine" for God knows what) and all kinds of industry. How will it affect the packaging industry?

Just a little background on Nanotechnology for those of you who are not familiar with it, do not read comic books or watch the Sci-Fi channel often. Nanoscience and Nanotechnology's primary push, excitement or "umph", if you will, is in the arenas of materials science, electronics, optoelectronics and biomedical. Biomedical is extremely exciting as it can be used hopefully for repairing organs, removing disease, or even making organs operate more efficiently. Then this will move into the military arena. How? A soldier can be repaired faster if injured. Or maybe the soldier can run faster if Nanotechnology or Micromachines are being used to enhance reflexes. The list goes on and on; you can use your imagination.

So Nanotechnology is the ability to manipulate molecules and atoms to create structures, which may be used in the real world. So how does this apply to packaging? Well the science will help deliver materials with greater functionality or durability for the increase in shelf life. Examples, that popcorn bag that junior tries to open will not spill over your new carpet; the coke can that can stand up to punctures a little better. How about that box in your garage that had the cement mix in it, which is now soaked with water? Having to deal with a mess like that would make anyone mad.

Nanotechnology will provide better fracture hardness for aluminum. Note the above example. Nanoscience will provide better tensile strength for items that use carbon fiber and improve flame resistance for plastics with "nanoclay" composites.

Gas-barrier characteristics of Nanoclay in food packaging have sparked great interest. Using Nanoclay in packaging films, i.e. shrink-wrap helps to create a better oxygen barrier. Nanoclay will also be used for the tracking of products through the supply chain. In the next 5 years 5 million pounds of nanocomposite material will go into rigid and flexible packaging. This will affect the packaging of soft drinks, beer, meats and a vast array of packaged foods and condiments. These will be the first consumable items to be packaged in Nanotechnology Packaging. One can certainly hope it will allow one to get that last bit of ketchup out of that little pack whilst one is sitting at the ballpark!!!

One of the problems with Nanotechnology as with any new technology is the cost. This is something new and many companies in the paper, packaging and printing industries cannot really see what Nanotechnology can do for them. Will the cost be outweighed by the benefit, hopefully?

If the affects of Nanotechnology and its challenges are not addressed then the future of compositeness amongst companies and organizations will be threatened.

The application of Nanotechnology shows a lot of promise in the world of packaging. It will vastly improve the packaging materials properties. The only problem is these improvements are years away. Will companies see the benefit of continued investment? Or will they decide not to reinvent the wheel. Who knows? Time will tell.

Technologist Eric Drexler envisioned a future in which machines far smaller than dust motes would construct everything from chairs to rocket engines, atom by atom; in which microscopic robots would heal human ills, cell by cell. Sixteen years after the publication of Drexler's book Engines of Creation, the molecular-scale technologies most immediately available to consumers are somewhat less fantastic: stain-resistant khakis and more durable tennis balls.

Much of the hype is gone from nanotechnology, the term Drexler popularized for his world of very small wonders. But something more interesting has crept in: sales. The khakis and tennis balls are bringing in money, as are dozens of other new products made and enhanced through nanotechnology. To be sure, most nanotech companies are still investing more in R. and D. than they are collecting in revenue. But many commercial applications are in advanced stages of development or already on sale: handheld devices that can sense anthrax spores, hand cream that can protect us from them and computer chips that are faster, cheaper and cooler (we're talking temperature here, not hipness) and retain data even when the power is shut off. Says Richard Smalley, a Rice University professor and Nobel-prizewinning chemist: "We are only beginning to see the things nanotechnology can do."

Nanotechnology takes its name from a nanometer (nm), a billionth of a meter, or about one one-hundred-thousandth the diameter of a human hair. In common usage, it refers to an array of new machines and materials whose key parts are smaller than 100 nanometers and to the new tools, such as Veeco Instruments' atomic-force microscopes and Nanometrics' inspection tools for semiconductor makers, that allow the tiny parts and particles to be observed and manipulated.

It is a mysterious realm in which the laws of classical physics yield to those of quantum mechanics, in which the powerful bonds between atoms overtake the effects of gravity that rule the big world. Yet scientists have moved beyond the basic exploration of nanotech to its exploitation. The National Science Foundation foresees a $1 trillion market by 2015 for nano products, and businesses and governments around the world are rushing to cash in.

The White House has proposed that $710 million be spent on nanotech research next year- a 17% increase over the 2002 budget- on the development of everything from water-filtration equipment to military uniforms made from "smart" materials that can guard against germ warfare. Governments in Asia and Europe are investing $2 billion in similar R. and D., according to CMP-Cientifica, a research firm in Madrid. "Nanotech is a three-legged race right now," says Mark Modzelewski, executive director of the NanoBusiness Alliance, based in New York City.

Even enthusiasts like Modzelewski caution that no one should expect an overnight nanotech revolution. The technology will evolve-"radically," he says- as its benefits seep into virtually every crevice of human industry, from toys to tanks. And even professional investors are cautious. "True venture capitalists are not investing. They are watching," says Glenn Fishbine, author of The Investor's Guide to Nanotechnology and Micromachines. Only a handful of "pure play" nanotech stocks exist, including Nanophase Technologies, in Romeoville, Ill., which makes nanoscale powders, among them zinc oxide particles for sunscreen that won't turn lifeguards' noses white. Still, investors in behemoths such as Intel, Samsung and Dupont have been indirectly funding nanotech development for years.

Five sectors surveyed here- consumer goods, computers, pharmaceuticals, energy and cars- deserve particular attention for the progress they have made toward bringing profitable products to market.

Tires and Toys

Physicist Harris Goldberg wants to revolutionize the $1 billion tire-sealant business, but until that goal is realized, he will settle for tennis balls. InMat, Goldberg's seven-employee company in Hillsborough, N.J., regularly ships to Wilson Sporting Goods 55-gal. drums filled with an environmentally safe liquid containing 1-nm-thick sheets of clay. When the material coats the inside of a tennis ball, it traps air far more effectively than standard rubber alone and doubles the life of the ball. Wilson's Double Core, which made its debut more than a year ago, sells at a premium in U.S. tennis shops and this year became the official ball of the Davis Cup competition.

InMat will take in just $250,000 this year, but Goldberg expects to double that figure in each of the next few years, largely on tennis-ball business. Meanwhile, he is working to convince tire manufacturers that by sealing their wheels with his technology instead of butyl rubber, the current sealant, they can produce tires that run cooler and safer, are lighter and increase a car's fuel efficiency. The U.S. Army has asked InMat to develop gloves that will protect soldiers from chemical agents. Goldberg's funding has come mostly from an angel investor and grants, which means he is still on the prowl for cash. "It's been a battle," he says. "It's still a battle. But we're looking at enormous growth prospects." A chemical process that adds "nano-whiskers" to cotton fabrics and renders them wrinkle and stain resistant explains new products from Eddie Bauer and Lee Jeans. The fabrics were developed by Nano-Tex, a Greensboro, N.C., company that is 51% owned by Burlington Industries, a textile firm that is struggling to emerge from bankruptcy. Nano-Tex has also developed active-wear fabrics that disperse and dry sweat. Later this year, it will launch a line, destined for socks, underwear and T shirts, that will channel body odors through the structure of the fibers.

Rod MacGregor, a high-tech entrepreneur, runs NanoMuscle, an Antioch, Calif., company that makes 3-in. motors suitable for everything from power windows to dolls with nuanced facial expressions. "I like to be on the wave of the next insanely great thing," he says. His motors work because the alloy nitinol can assume different shapes as its temperature fluctuates. An electrical current causes a nitinol wire in the device to shorten, allowing the linear motor to contract like a human muscle but at 1,000 times the strength.

That's a simple task but an important one, and one MacGregor believes can reach markets worth $3.8 billion.

The NanoMuscle, which costs less than $1 to make, qualifies as nanotech, the company says, because of the size of its nitinol crystals, not the wire or motion. MacGregor compares his product to $40-to-$100 small motors made by potential competitor RMB, of Biel-Bienne, Switzerland. Hasbro, a major investor in MacGregor's start-up, expects to deliver its first nano-powered toys by Christmas 2003. NanoMuscle's challenge, like InMat's, will be to stay afloat long enough to sign companies on as clients.

Smart Soot

Like MacGregor, Greg Schmergel is a serial entrepreneur. If his company, Nantero, in Woburn, Mass., is successful, it will eventually add about five minutes to everyone's day. The company wants to build an "instant-on" computer that doesn't need to boot up.

In a year, Nantero expects to produce a commercial prototype for a chip with "nonvolatile random-access memory" (NRAM), which means its chips won't forget how to run all its programs when the power is switched off. The technology uses arrays of 2-nm strands of carbon atoms, called carbon nanotubes, that convey electrons faster than copper and are 100 times as strong as steel at a fraction of the weight. Pairs of tubes store data by locking together when a current runs through them and stay together even when the computer power is switched off and back on.

The tubes remain linked until separated by a countercurrent, so their memory is retained. And these chips have other advantages. Schmergel says that within three years, Nantero can bring to market chips with nram that can store 10 times as much data as a silicon chip the same size while operating faster and with less heat. "They're not saying much publicly about their approach," says Steven Glapa, president of the nano-consulting firm In Realis, "but what they're promising sounds pretty breathtaking."

Nanotubes could be the first commodity in the nanotech economy. Dozens of companies around the world already pump out mounds of the stuff- affectionately called soot- and sell it to some of the world's largest companies and labs for research: IBM, Hewlett-Packard, Samsung and NEC. Nano-Lab, in Brighton, Mass., is one of the few nanotech companies turning a profit. It sold $200,000 worth of made-to-order nanotubes in 2001 and is on track to more than double that amount this year. Last week HP researchers unveiled a way of manufacturing molecular-scale circuitry that will be cheaper and use less power than current silicon chips and have the potential to store entire libraries of information.

Cancer Busters

Professional athletes won't find Richard Smalley's soccer balls quite as novel as Harris Goldberg's tennis balls. In fact, without an atomic-force microscope, they won't find them at all: the naturally occurring structures are composed of just 60 carbon atoms. Yet Smalley's discovery is expected to help treat aids, cancer and Lou Gehrig's disease, and it earned him and two colleagues the 1996 Nobel Prize for Chemistry.

These molecular structures are called fullerenes, or buckyballs, in honor of the American architect and inventor Buckminster Fuller. Smalley sits on the board of C-Sixty, a biotech company that builds fullerenes into molecules that researchers hope will attach to and deactivate hiv molecules and blow up cancer cells on cue. "Buckyballs are not quite like nanosubmarines that target deadly diseases"-as seen in the 1966 film Fantastic Voyage-"but because of their size and shape, they are well suited for drug discovery," says Stephen Wilson, co-founder of C-Sixty, based in Houston.

C-Sixty's hypothesis is that buckyballs offer a master-key approach, functioning as a universal molecule that can be, in a sense, weaponized to attack any enzyme or receptor that plays a role in a disease's development. C-Sixty is assembling libraries of new buckyball-based molecules that it will test for potential therapeutic value. Early next year, it will conduct human trials of fullerene-based drugs for hiv and Lou Gehrig's disease. With about one-tenth the toxicity of the current hiv drug cocktails, the company's molecule targets new strains of the constantly mutating virus that are no longer susceptible to treatment. In the case of Lou Gehrig's disease, a degenerative nerve illness, the drug prevents or repairs neurological damage. C-Sixty has no revenue yet but will soon announce a partnership with "one of the top three pharmaceutical companies," says its president, Uri Sagman.

Other researchers are building nanoinventions so small that they can slip inside diseased cells and halt their development. Dr. James Baker, head of the University of Michigan's Center for Biologic Nanotechnology, is developing nanoscale molecules called dendrimers to target cancer cells. He says that "nanotechnology gives us a totally new set of tools to diagnose and treat disease," including an ability to eliminate cells before they become cancerous.

Baker and his team created a company called NanoBio. An $11 million Pentagon grant allowed the team to develop a cream that can penetrate and kill infectious microbes, everything from the fungus that grows on toenails to flu viruses to anthrax spores. The military version, called NanoDefend, is a liquid designed to decontaminate clothing and surfaces that have come into contact with anthrax, Ebola or smallpox. A creamy gel or goop, called NanoGreen, can be used by the military to decontaminate skin- and may eventually have topical and vaginal applications for consumers, according to NanoBio ceo Ted Annis. The firm, which hopes to partner with existing companies, is preparing to submit seven products to the fda for approval, a process that takes several years.

Some companies will simply help common drugs work more efficiently. Elan Drug Delivery, located in King of Prussia, Pa., pulverizes existing drugs to a size that maximizes the body's ability to absorb them. Naproxen sodium, a pain medication found in products such as Aleve, can take as long as two hours to exert its pain-relieving effect. Nano Systems has developed a crystal version of naproxen, still in clinical development, that works in 15 to 20 minutes. "Using NanoCrystals has not made naproxen a better drug"-just seven to eight times as fast as the commercial product, says Larry Sternson, president of drug delivery at Elan, the Dublin-based drugmaker that is the parent company of Elan Drug Delivery. Detection and analysis are also enhanced by small technology that is not strictly nano-scale. MesoSystems, a young but profitable firm, sells to fire departments handheld devices that collect biological particles 0.5 to 10 microns across- anthrax, for one- and preserve them in a liquid for identification. MesoSystems supplies Lockheed Martin with an air sampler it uses in its Biomail Solutions product, a biohazard detector in field testing at some federal agencies. MesoSystems made about $250,000 last year on revenues of $7 million and this year hopes to gross more than $10 million.

THE CLEANER ENERGY

As an alternative to fossil fuel, everyone loves hydrogen fuel cells, which produce clean energy out of hydrogen and oxygen. But hydrogen, while abundant in the air, isn't widely available in refined form. And machines that run on hydrogen are equally scarce. Researchers at the Tokyo Institute of Technology have been working on the first problem, automakers on the second. The Tokyo group has developed a way to "crack" hydrogen, using a mesh of thin carbon fibers studded with molecules of a nickel compound. The filter breaks down natural gas into carbon and hydrogen that is pure enough for use in fuel cells.

Another impediment is the cost and supply of the platinum particles that catalyze, or kick off, the process. Think of them almost as matchmakers, encouraging every oxygen atom to mate with two hydrogens, releasing valuable energy with each reaction. That is the heart of the fuel cell.

Because of the current size of these catalyst particles, about 10 nm, and their tendency to clump together, platinum is not used efficiently. The world's entire annual output of platinum would not meet the demand if fuel cells were used by only 10% of cars produced worldwide. Hydrocarbon Technologies- which is owned by Headwaters, an alternative-energy company based in Draper, Utah- says it has found a way to create nanoscale platinum particles that won't clump together and slow down the process, as current ones do. The new particles are expected to keep fuel cells running in a stable, efficient manner and stretch the platinum supply. Tim Harper, founder of CMP-Cientifica, says these particles show how "nanotechnology can make previously uneconomic processes viable" for businesses.

OPEC has more than fuel cells to worry about from nanotechnology. Last month China's largest coal company licensed U.S. technology that will enable it to build a $2 billion coal-liquefaction plant in Inner Mongolia. The heart of this new technology is a gel-based nanoscale catalyst that improves the efficiency of coal conversion and reduces the cost of producing clean transportation fuels. If the technology lives up to its promise and can economically transform coal into diesel fuel and gasoline, coal-rich countries such as the U.S., China and Germany could depend far less on imported oil. At the same time, acid-rain pollution would be reduced because the liquefaction strips coal of harmful sulfur. Given current world oil prices (about $27 per bbl.), turning coal into gas is economical in China. "A $4-to-$8-per-bbl. increase in the price of oil would make it economically attractive in the U.S. too," says Theo Lee, ceo of Hydrocarbon Technologies.

PLASTIC CARS

Every big carmaker promises that any year now, it will have a fuel-cell car on the road- a vehicle that will cruise silently, spit drinkable water from its tail pipe and provide power to your house when you plug it into the garage. In the meantime, auto manufacturers are putting nanotechnology to work in other ways. Toyota was the first to experiment with strong, lightweight nanocomposite materials in the late 1980s, and U.S. automakers are starting to move nanocomposites out of the lab and into vehicles. General Motors is using advanced plastics to make step assists for 2002 GMC Safari and Chevrolet Astro vans.

The new materials are stiffer, lighter and less brittle in cold temperatures than other plastics. Improvements in strength and reductions in weight lead to fuel savings. The next step is for GM to use nanocomposites in car interiors and bumpers and eventually in load-bearing structural parts, such as vehicle frames. As nanotechnology produces more products and processes, will the technology ever catch up with Eric Drexler's theories? Says Steve Bent, a Washington patent lawyer for nanotech firms: "That will be the research agenda for the rest of the century."

TRADE TOOLS

A primary long-term goal of nanotechnology research now underway worldwide is the creation of micro machines-mobile, versatile devices and systems that can perform engineering tasks or medical procedures on a microscopic scale.

Thanks to their minuscule size and motility, microdevices will open up new frontiers in molecular materials and manufacturing, electronics, health care, medical research, biotechnology, and information technology.

Before micromachines become a reality, however, researchers must learn a lot more about the properties and interactions of materials in the micro- world, where forces other than gravity dominate. They'll also have to develop the tools and technology for building micromachines, which are the focus of research efforts underway at the University's Advanced Microsystems Laboratory.

Researchers there already have developed smart microassembly systems, atomic-level force sensors, and various microdevices to aid the development of micromachines, says Bradley Nelson, associate professor of mechanical engineering.

Some of these parts are no larger than one one-hundredth of the width of a human hair, says Nelson, who investigates the physical interactions among micromechanical parts during the assembly process. Lab researchers had to build special sensors that could measure nanonewtons of surface-effect forces over nanometers of separation between parts.

"We're working to develop a new manufacturing paradigm so these parts can be assembled economically into intelligent microscopic machines," he explains.

Micromechatronic systems span a broad range of applications, from microsurgical tools for minimally invasive surgery to machines that can manipulate bits of information and improve disk-drive density to "clouds of intelligent dust" that monitor the environment. Other applications for Nelson's research include the automatic assembly, computer vision, manufacturing, and robotics industries.

Another University researcher is studying active thin films now being developed for small-scale sensors and micromachines. Mathematics professor Mitchell Luskin is collaborating with colleagues from several IT departments to understand the behavior of thin films of martensitic, ferroelectric, and magnetostrictive materials.

Luskin is working to apply this research to the development of microelectromechanical systems (MEMS) that can have an important impact on biomedicine.

"A transformation is just beginning in biomedicine that promises to revolutionize how some surgeries and therapies are done," Luskin says. MEMS may someday replace the knife and scalpel as tools for minimally invasive surgery and also deliver powerful drugs directly to targeted areas in the body.

According to Nelson, the University has emerged as a national leader in the research and development of nanotechnology and micromachines. "The University is a real player in this field," he says.

5

FUTURE OF NANOTECHNOLOGY

It may seem like science fiction, but the nanotechnology revolution is already underway. It's going to change our lives in ways we've only begun to imagine.

Nearly 40 years ago, the late Isaac Asimov wrote a sci-fi classic about a dramatic race against time and a technology so unbelievable that the author aptly titled his book Fantastic Voyage. Back in 1966, Asimov's scenario was sensational. In retrospect, however, it's his prescience that's amazing.

The book and the film it inspired tell the story of a desperate struggle between Cold War adversaries for control of a still-experimental technology called miniaturization, which reduces the size of very large objects by shrinking their atoms. A five-member medical crew undergoes extreme miniaturization and travels through the body of a professor (and communist defector) via a tiny submarine. The crew's mission is to destroy a life-threatening blood clot in the defector's brain, thereby saving his life and the intellectual capital that will ensure democracy's triumph.

Today, the technology Asimov envisioned in Fantastic Voyage is far closer to becoming fact than many people realize. A burgeoning new field called nanotechnology has sparked the scientific imagination and generated a surge of research activity across many disciplines. It's also instigated a high-stakes race for knowledge and dominance of the still-experimental technology. Many experts believe that nanotechnology will someday revolutionize science, medicine, and industry. Nanotechnology manipulates and controls the physical world's smallest, most basic components to obtain a desired result. The properties of a material-such as its mechanical strength, electrical conductivity, and complexity-are determined at the molecular level and are dependent upon the arrangement of its atoms. If they had the ability to manipulate and control the structure of materials at or near the scale of a nanometer-a billionth of a meter-researchers theoretically could invent and improve materials, technologies, and products to a degree that's almost unimaginable today. The most basic example of existing nanotechnology is the nanoscale molecular machinery inside the living cell. Each cell works like a machine, converting fuel to energy and pumping out proteins and enzymes as directed by its DNA. However, within the limitations of its existing structure, a cell can perform only certain highly circumscribed functions. Nanotechnology will break down these limitations and give humans unprecedented control over the material world. Nanotechnology's ultimate goal is the creation of ultrasmall computers and machines as well as advanced materials that can carry out vital engineering or medical tasks at the molecular level.

Advances in nanotechnology will affect a number of major industries, including electronics, medical science, pharmaceuticals, and computers. It could lead to better recording devices, faster computers, and better protective coatings.

Researchers also envision some amazing products and devices that smack of science fiction, such as an "invisible" airplane with a chameleon-like ability to blend into its surrounding environment, thanks to a special exterior coating. Someday construction materials may come equipped with built-in sensors that detect weather conditions and control the indoor climate of homes and offices automatically.

And a technology that was the stuff of science fiction just a few decades ago is a step closer to becoming reality at the University. Researchers here are working to create a gene gun, a nanoscale medical weapon that can enter the human body and fight diseased cells one at a time. The project is one of the University's most important nanotechnology accomplishments to date.

"The 'U' has pockets of excellence in this field," says David Pui, Distinguished McKnight University Professor and director of the University's Particle Technology Laboratory, where the experimental gene gun is being developed. "Nanotechnology is inherently multidisciplinary, and a comprehensive university like ours provides excellent opportunities to conduct such research."

Still in its infancy, nanotechnology has captured the attention of researchers and world governments alike. However, when physicist Richard Feynman introduced the concept of nanotechnology in 1959, the idea of manipulating matter atom by atom received a cold reception. Serious research in the field didn't begin until the late 1980s and has only recently become a primary focus of researchers.

National governments have accelerated their support for nanotechnology research and development. In 1999, the U.S. government invested about $260 million in nanotechnology research, and Japan and Europe made similar investments. Last January, President Clinton declared a National Nanotechnology Initiative, and his FY 2001 budget request included a $227 million increase (84 percent) in government funding for research and development. The White House National Science and Technology Council recently created the Interagency Working Group on Nanoscience, Engineering, and Technology. Members from eight federal agencies are working closely with academic and industry leaders to explore nanotechnology's potential on a national level.

The National Science Foundation (NSF) currently is receiving proposals to fund nanotechnology research. Years of experimentation and research in nanotechnology have yielded promising results for the University, which is preparing a proposal for submission to the NSF. University researchers are working on nanotechnology projects in micromachines, medical devices, computers, electronics, and a number of other fields. Others are exploring the world of microsystems and particle interaction.

All this research activity has moved the University to the forefront of the nanotechnology revolution, according to industry observers. The University also has won recognition for promoting interdisciplinary research collaborations among various academic units. The Institute of Technology, College of Biological Sciences, Academic Health Center, and College of Agricultural, Food and Environmental Sciences all are involved in nanotechnology experimentation and research.

For the hundreds of medical device, telecommunication, manufacturing, and biotechnology companies based in Minnesota, advances in nanotechnology hold the key to future innovation and economic success. Last March the University hosted the Minnesota Nanotechnology Summit, which brought national experts together to discuss the impact and potential of nanotechnology. The summit also gave the University a chance to showcase its accomplishments.

According to University president Mark Yudof, the conference was a first step toward identifying how Minnesota's public and private sectors, together with the University, can capitalize on nanotechnology's potential and guarantee that the University remains among the leaders in the field.

"There are a lot of good things going on at the University of Minnesota," Pui says. "The nanotechnology summit was one of the first among research universities nationwide to promote interdisciplinary collaboration. The University's central administration and the deans are very supportive of this initiative."

On the cusp of a technological sea change, the world is likely to see a number of scientific breakthroughs in nanotechnology over the next 25 years. University researchers are working diligently to ensure that they're responsible for some of those fantastic journeys.

FUNDAMENTALS AND THEIR EXPLORATION

High hopes and unanswered questions accompany any emerging technology, and nanotechnology is no exception. For nanotechnology to fulfill its promise, researchers must find the answers to fundamental questions about how microsystems and particles interact with each other.

To fill that knowledge gap, some researchers are turning to mesophysics, part of the broader interdisciplinary field of nanoscience. The mesoscopic scale encompasses systems that are larger than single atoms but small enough that their properties differ radically from larger pieces of matter. An understanding of mesophysics is essential to achieving higher densities of microelectronic semiconductor-based circuits.

One such project is the focus of Leonid Glazman, McKnight Presidential Chair of Theoretical Condensed Matter Physics, who is considered one of the world's leading authorities in the field. Glazman's research explores the theory of electron transport and correlations in nanoscale devices.

In most conductors, electrons carry the electrical current through a wire, like a stream coursing through its channel. The width of a conventional wire is about ten million times the size of an electron, so the electron can pass through it with ease.

The quality of a wire as a conductor of electrons is defined by the wire's conductance. When wires become extremely small, however, unusual phenomena occur. Conductance breaks down when a wire's width is small enough to guide or control the passage of single electrons through the wire. Then the quantum properties of electrons- their ability to behave as waves, as well as the finite and universal value of electron charge-become important.

Thanks to advances in the semiconductor industry, it's now possible to produce small devices that morph continuously, ranging from a waveguide for quantum waves of electrons to islands that carry a tiny discrete charge produced by only a few electrons.

By studying the fundamental properties of these smaller mesoscopic systems, including the particle and wave properties of electrons, Glazman and other physicists are uncovering the fundamental physics behind nanotechnology.

Other University researchers are intent on learning how to control the size, composition, and other properties of nanoparticles.

Andreas Stein, an associate professor of chemistry, is developing nanoporous solids that mimic the structures of natural porous minerals called zeolites. Functioning as molecular sieves, zeolites allow molecules of designated sizes and shapes to pass through and react inside their channels under very controlled conditions. Synthetic zeolites have various applications in industry. The petroleum industry uses them to increase the octane rating of gasoline, a mixture that contains molecules of various sizes. When gasoline is filtered through zeolites that have reactive channel walls, certain molecules in the gasoline react with the walls in a process that increases the octane levels. For the reaction to occur, however, the molecules must come in contact with the channel walls.

One aspect of Stein's research has dealt with modifying the surface reactivity of porous solids. "We want to go beyond merely replicating nature's three-dimensional zeolites," he says.

Typically, the channels or pores of natural zeolites are just large enough to accommodate small molecules, but researchers need porous solids that give them more flexibility and control. If researchers can vary the size and arrangement of the channels and pores, they'll have greater control over the types of molecules that can interact with the sieve-and over other properties of a given material.

It's already possible to design synthetic porous solids to accommodate larger molecules. "We can now prepare solids with multiple pore sizes and in a large variety of compositions," Stein says. "This [capability] has opened the door to new applications."

Other possible applications for modified porous solids include drug delivery, catalysis, and optical technology.

One of Stein's more recent projects involves the synthesis of photonic crystals, which don't absorb light of a given color. If a paint were created using photonic crystals that don't absorb the color green-the most intense color in the spectrum-then a car whose exterior was coated with that paint wouldn't absorb green light and would remain cooler when exposed to direct sunlight.

In the Reacting Flow and Nanoparticle Laboratory, Michael Zachariah is studying methods of controlling the morphology of nanoparticles grown from the vapor phase. "The idea, essentially, is that the properties of particles sometimes depend on their size," he says. "If you can control the size, you can control the property."

Zachariah, an associate professor of mechanical engineering and a member of the chemistry department's graduate faculty, says that the ability to manipulate particles would benefit nearly every manufacturing industry, especially in the areas of ceramics, electronics, and fiber optics.

Although they're exceedingly small, nanoparticles can contribute to problems on a global scale.

Peter McMurry, professor and head of the mechanical engineering department, has been studying atmospheric nucleation since 1972. Gas-phase chemical reactions cause atmospheric nucleation-the formation of particles of precipitate-which affects the earth's radiation balance and the formation of aerosol-induced haze.

Over the years, McMurry and his research team have made several important discoveries and developed instruments to detect the presence and measure the composition of nanometer-sized particles. One of these instruments has been commercialized and is now widely used. McMurry's group has conducted field measurements of atmospheric nucleation over the Arctic Ocean, Hawaii, Colorado, Atlanta, the Indian Ocean south of Tasmania, and at the South Pole.

Steven Girshick, professor of mechanical engineering, is the principal investigator of a study that tests nanostructured films for friction and wear-resistance. The three-year, $660,000 project is funded by the National Science Foundation. His research has led to new methods of depositing stronger and more efficient nanostructured coatings.

But depositing the nanoparticles can be tricky. As a flow carrying particles approaches a surface, very small particles follow the flow as it curves around the surface, like snowflakes that curve around a car's windshield. To deposit a coating of nanoparticles efficiently, the particles must be traveling at high velocity.

Girshick and his colleagues invented a patented process called hypersonic plasma particle deposition, in which nanoparticles nucleate in a plasma that's accelerated through a nozzle. The particles-about 20 nanometers in size-hit the targeted surface at a velocity of two kilometers per second, creating a dense nanocrystalline coating. According to Girshick, this type of coating would improve friction and wear-resistance in automotive shafts and ball bearings as well as in biomedical implants. "There is simply a huge number of possibilities," he says. "There's been a lot of work [being done] worldwide to make nanostructured coatings. This process is unique to Minnesota. It is a different approach."

THE CONCEPT OF NANOELECTRONICS

Experts everywhere say that nanotechnology will lead to smarter, faster, and more efficient electronics, but they also admit that researchers face some challenging obstacles along the information superhighway.

Particle contamination is one of the thorniest problems facing the electronics and computing industries. Even the tiniest particle can destroy a computer chip and cause other damage. Several University researchers are searching for ways to detect particle contaminants and prevent their formation.

A team of researchers headed by mechanical engineering professor Peter McMurry is studying the formation and growth of contaminants in semiconductor processing equipment. Smaller contaminated particles deposit more quickly than larger ones and can destroy a number of memory chips and processors. The team's goal is to design new processes or tools for producing contaminant-free devices.

"It's important to understand how to prevent particle deposition, and this will require an understanding of how such particles are formed and transported," McMurry says.

His team has developed a device called the particle beam mass spectrometer, the only available instrument for measuring nanoparticles in low-pressure semiconductor processing equipment.

Electrical engineering professor Steve Campbell is among the researchers who are looking for ways to create electronic devices with greater memory. His studies are aimed at detecting and preventing the formation of particle contaminants in chemical reactors. Although research in the area is still relatively new, some early prototypes have been made, says Campbell.

Campbell's Microtechnology Laboratory team has developed a machine to detect and measure particles. "Now we're trying to demonstrate that isolating these particles and using them to our advantage is feasible," he says, adding that his group is seeking funding to continue its research.

"The importance of smaller and faster semiconductor devices is evident from the trends we've all seen [toward] faster, more powerful, and less costly computers," says McMurry. "Work that is being done to fabricate devices at the nanoscale will ensure that this trend will continue for some time to come."

The same is true in recording media. Associate Professor Randall Victora and Professor Jack Judy of the electrical and computer engineering department received grants two years ago from the National Storage Industry Consortium and Seagate Corporation to develop superlattice magnetic recording media.

The focus of their research is magnetic recording technology for hard drives, which are vulnerable to thermal fluctuations. These fluctuations can affect the recording process and destroy vital information stored on a disk in the drive.

"We're preventing that from happening," Victora says. "These superlattices cause the disk to have a stronger resistance to these thermal fluctuations."

Once developed, the technology would be used primarily by the computing industry. Companies such as Quantum, IBM, and Seagate already have expressed interest in the group's research, Victora says.

Other University projects also have implications for the electronics industry, including the study of adhesion in computer chip manufacturing, molecular electronics, and the search for an appropriate nonconductor to replace copper, which is limited in its ability to transport information between resistors.

DNA Scaffolding

The next frontier for information processing may lie at the interface of nanoelectronics and biotechnology.

An interdisciplinary team led by electrical and computer engineering professor Richard Kiehl is exploring the use of DNA as a programmable scaffolding for the self-assembly of nanoscale electronic components. As a model for fabricating and designing semiconductor devices and circuits, DNA offers two key advantages: size scale and programmability.

Most industry experts believe that within the next 10 to 15 years the ability to scale down conventional technologies will reach its limit. At that point, the operating principles of conventional devices-and the techniques used to fabricate them-will break down. The basic elements of the DNA molecule are at just the right scale, says Kiehl. Self-assembly uses bio-recognition, a natural process in which one molecule is attracted to and binds with another to form small structures. In the case of DNA, the attraction can be programmed so that the molecules will spontaneously assemble in solution to achieve a desired result.

"It's possible to synthesize small versions of DNA molecules in the laboratory and program in whatever code you want," says Kiehl. "And because the two strands of DNA have complementary codes that match up, you can design one strand of DNA in a certain way so it will match another strand and assemble a nanoscale structure this way."

The matched segments form a scaffolding on which nanoparticles are affixed at highly selective attachment points. It's an approach that offers the programmability and precision needed for assembling electronic circuitry on the nanoscale.

"We have to make a real paradigm shift," Kiehl says. "Not only do we have to keep improving performance, but we also must look at the kinds of devices we can make at those scales and how we want to use them to process information."

To that end, the researchers are turning to the human brain for inspiration. They envision devices whose electrical characteristics resemble those of neuron-like electrical waveforms in the brain. Like certain regions of the brain, the devices would process information based on pattern recognition rather than on individual bits of information. It's a more sophisticated level of information processing than can be achieved using conventional computers. Kiehl predicts there will be a wide range of applications for this technology, including signal processing, communications systems, and computer systems. "The higher end of this [work] will be things that computers can't do very well today because the operations they use are too restrictive. One is the ability to recognize a pattern, such as identifying a letter as being an 'A' or a 'B', or being able to identify a face.

"It won't be just making things faster and faster in the conventional way," he says. "It will really be opening up new ways to process information in machines."

The Gene Gun

David Pui has big dreams, all generated by an ultrasmall technology, and he's not alone. Leaders from industry, academia, and government, including President Bill Clinton and University president Mark Yudof, agree that nanotechnology could someday revolutionize science, medicine, and industry.

Pui, a Distinguished McKnight University Professor in mechanical engineering, is director of the Particle Technology Laboratory-a hub of nanotechnology research at the University-and a staunch advocate of the emerging field. Pui, who chaired the Minnesota Nanotechnology Summit last March, says that the University has already established itself as a leader in promoting and supporting interdisciplinary nanotechnology research. Pui should know. He's been involved in one of the University's most promising nanotechnology achievements to date, the development of a "gene gun"-a newly patented electrospraying apparatus for continuous gene transfection. The gun shows great potential as a delivery method for gene therapy.

Nowhere are the hopes for nanotechnology riding higher than among medical researchers. The gene gun project is the result of years of research by Pui and Assistant Professor Da-Ren Chen, also a member of the mechanical engineering faculty. Christine Wendt, an assistant professor in the Medical School who has worked with Pui's team, says she can readily foresee ways in which the gun could have medical applications. The goal of gene therapy is to treat cells with genes that will help the cells produce specific proteins to fight disease. But in order for gene therapy to work, a number of things must happen. The gene first must get to the cell, and then the cell must accept it. Finally, the gene must arrive at the nucleus for its effects to take hold.

Traditional methods for this process, called gene transfection, include uptaking plasmid DNA, infecting the cell with viruses that contain the desired gene, fusing cells together, and microinjecting cells. All these methods treat a large number of cells simultaneously, but they tend to have a low efficiency rate. In treating lung cancer, for example, doctors have been frustrated by their inability to successfully deliver genes to diseased cells.

"The traditional ways of treating many diseases have limitations, but gene delivery offers new and exciting possibilities," Wendt says. "When you use chemicals to insert genes into cells, there are efficiency and price issues. Shocking the cell with electricity can cause serious damage, and it's a method that's limited to cells contained in petri dishes. The important thing is to optimize gene transfer carefully, which this gene gun can do. It has the potential to be applied directly to the treatment of human disease."

The physical transfer of genes is not a new concept; however, the technology behind the University's gene transfector is novel and offers distinct advantages over its precursors.

In the late 1980s, horticultural scientist John Sanford and his colleagues at Cornell University designed a gene gun that has been widely used, most commonly in the genetic transformation of plants. To deliver genes to the targeted cells, the Sanford gun uses a bullet-like projectile coated with gold particles that act as gene carriers.

The gun itself consists of a highly pressurized upper chamber and a low-pressure lower chamber with a diaphragm in the middle. When the diaphragm is punctured, pressure from the upper chamber emits a shock wave that hits the projectile, discharging it forward until the projectile hits a porous screen. The force of the impact launches the gold particles through the screen downstream toward the targeted cells. The particles hit the cells at a velocity high enough to puncture the cell membrane, penetrate the cells, and release the genes, which are diffused into the nucleus.

According to the University researchers, the Sanford gun has a number of disadvantages. The individual gold particles that coat the projectile are harmless, but they have a tendency to stick together and crush the cell, an effect called "pit damage."

Sometimes researchers have to shoot the same cell sample repeatedly to obtain the desired result. After each shot, the diaphragm and projectile must be reloaded and the process repeated. If there are thousands of cells to shoot, the preparation of the gene-coated carriers becomes a grueling, time-consuming process. Although it's similar in concept to the Sanford device, the University's gun uses a very different technology, a patented process called continuous gene transfection. A gene suspension-with or without gold particles-is loaded into a capillary via a syringe. The suspension is sprayed out of the capillary at a controlled pace under a high electric field, which is created by applying voltage. The spray itself consists of highly charged and dispersed gene particles or gene-coated particles. Particles with a similar charge repel each other at a velocity high enough to penetrate the cell membrane and release the genes into the cells.

"It's a simple process," says Chen, who is credited with designing the electrospray process. "But it's just how you engineer it to get it to do what you want it to do."

The University researchers also had to be careful not to infringe upon the Sanford gun's patent. Specifically, they had to avoid a technology that used external forces to discharge the gene carriers into the cells, as the Sanford gun does. Pui and his team had to devise a different technique. Furthermore, they had to design the gun with a specific application in mind. That issue was resolved after Wendt met Pui and Chen, who discussed their ideas for the gene gun project with her. Wendt, who was working on another project involving the effects of small particles, says she knew instantly that the gene gun had great potential for application in her field.

She worked closely with Pui and Chen in testing the gun. In one experiment, the researchers injected enhanced green fluorescent proteins (EGFP) from jellyfish into different types of cells, including African green monkey cells and human lung-cancer cells. The EGFP proteins glow when exposed to ultraviolet light, so the researchers could see whether or not the genes had entered the cells.

Experiments also proved that the gun's design and technology offered other advantages over existing methods of gene transfection. The charged particles repel each other with such force that there's no risk of pit damage. The gun's capillary-and-syringe setup permits continuous injection of the genes. In fact, Pui says, the process could run continuously for days before the gene suspension had to be reloaded. It's a method that could be adapted to a system of mass production, in which the target cells or samples are placed on a conveyor belt and transported to the spraying assembly.

"We can shoot until it's sufficient," Chen says. "[With the Sanford gun], they have to shoot sometimes five times, sometimes up to 30 times, and that can really be a headache. Plus it takes a while to prepare the gene formula on the projectile. When they have to change, it is a lengthy process."

Because their gene gun worked successfully on different types of cells, Pui and his team believe that it can be used on an even broader range of cell types. The experiments also proved that the gold particles weren't essential to the technology and that the process could be conducted on a mass scale.

The years of design and experimentation have paid off. On July 25, Pui and Chen received U.S. patent No. 6,093,557 for an "electrospraying apparatus and method for introducing material into cells." It's one of three patents Pui and Chen have received for their nanoparticle research over the past two years. The other patents, issued last fall, are for devices used to charge the particles. Another patent request is in the works, but the researchers are still cautious about discussing it.

"We've demonstrated that the devices work," Pui says. "They have been proven, and because of that, outside interest has been high." He says one local company has licensed the nanoparticle charger patents and another large Minnesota company has expressed interest in the gene gun-related devices.

Although test results so far have been very successful, the gun is still in the earliest stages of development. Pui and his team have identified several major goals for the next phase of their research, which will subject the device to more difficult challenges. Besides improving the gun's design and dosage controls, they intend to use it on more-difficult cell types and experiment with different genes.

The team also wants to expand the range of the gun's applications by making it portable. Theoretically, doctors treating lung cancer could insert a portable gun into the patient's lung and spray the diseased cells. Once inside the lung, the gun would release the gene, which the patient then could inhale.

"We need to continue to work on this gene gun project," Pui says. "Furthermore, we need to continue to work with other departments, like we did with Christine (Wendt). This work crosses many fields."

It's work that requires both the visionary's capacity for big dreams and the scientist's meticulous, ceaseless scrutiny. But above all, it's the dreaming that compels researchers like Pui, Chen, and Wendt to probe deeper into the tiny world of nanotechnology. And their research has already given them a taste of what lies ahead.

As Chen says, "Now we have to explore all of the possibilities."

6

NEARING THE NANOTECHNOLOGY REVOLUTION

There's a lot of buzz-nanotechnology is "coming soon." But what is nanotechnology? Why doesn't anyone ever explain that? Well, it's not that easy. While experts agree about the size of nanotechnology-that it's smaller than a nanometer (that's one billionth of a meter) they disagree about what should be called nanotechnology and what should not. Looking back at the historical roots of nanotechnology helps us get a better grasp on what nanotechnology is and why it's important now, and how it will change the world in the future.

The story of nanotechnology begins in the 1950s and 1960s, when most engineers were thinking big, not small. This was the era of big cars, big atomic bombs, big jets, and big plans for sending people into outer space. Huge skyscrapers, like the World Trade Center, (completed in 1970) were built in the major cities of the world. The world's largest oil tankers, cruise ships, bridges, interstate highways, and electric power plants are all products of this era. Other researchers, however, focused on making things small. In the 1950s and 1960s the electronics industry began its ongoing love affair with making things smaller. The invention of the transistor in 1947 and the first integrated circuit (IC) in 1959 launched an era of electronics miniaturization. Somewhat ironically, it was these small devices that made large devices, like spaceships, possible. For the next few decades, as computing application and demand grew, transistors and ICs shrank, so that by the 1980s engineers already predicted a limit to this miniaturization and began looking for an entirely new approach.

As electronics engineers focused on making things smaller, engineers and scientists from an array of other fields turned their focus to small things-atoms and molecules. After successfully splitting the atom in the years before World War II, physicists struggled to understand more about the particles from which atoms are made, and the forces that bind them together. At the same time, chemists worked to combine atoms into new kinds of molecules, and had great success converting the complex molecules of petroleum into all sorts of useful plastics. Meanwhile geneticists discovered that genetic information is stored in our cells on long, complex molecules called DNA (about 2 meters of DNA is packed into each cell!) This and other work led to a greater understanding of molecules, which, by the 1980s, suggested entirely new lines of engineering research.

So, the roots of nanotechnology lie in the merging of three lines of thinking-atomic physics, chemistry, and electronics. Only in the 1980s did this new field of study get a name-nanotechnology. This new name was popularized by physicist K. Eric Drexler, who pointed out that nanotechnology had been predicted much earlier, in an almost-forgotten 1959 lecture by Nobel Laureate Richard Feynman, who proposed the idea of building machines and mechanical devices out of individual atoms.

The resulting machines would actually be artificial molecules, built atom by atom. While the resulting molecule might itself be larger than a nanometer, it was the idea of manipulating things at the atomic level that was the essence of nanotechnology. But not only was this kind of manipulation impossible at the time, but few people had any idea why it would be useful to do it! With all the new research, however, Drexler revived Feynman's vision and helped introduce the general public to the basic concepts of nanotechnology.

Although nanotechnology dates from the 1950s, the biggest changes have occurred just in the past few years. In the late 1990s, research money began pouring in from corporate and government sources. In the space of just a few years governments around the world launched three major (and many other smaller) new research programs, including the National Nanotechnology Initiative in the U.S. and the nanotechnology branch of the European Research Area. Japan has its own huge nanotechnology program, with money coming from private industry and government agencies such as the Ministry of Trade and Industry.

THE BUILDING BLOCKS

Nanotechnology is a field that's just being established, and although there are big plans for the smallest of technologies, right now, most of what nanotechnologists have accomplished falls into three categories: new materials-usually chemicals-made by assembling atoms in new ways; new tools to make those materials; and the beginnings of tiny molecular machines.

Some of the primary building blocks in nanotechnology are buckminsterfullerenes (almost always known as buckyballs or fullerenes), which are clumps of molecules that look like soccer balls. In 1984 Richard Smalley,

Robert Curl, and Harold Kroto were investigating an amazing molecule consisting of 60 linked atoms of carbon. Smalley worked these atoms into shapes he called "fullerenes," a name based on architect Buckminster Fuller's "geodesic" domes of the 1930s and first suggested by Japan's Eiji Osawa. Sumio Iijima, Smalley, and others found similar structures in the form of tubes, and found that fullerenes had unique chemical and electrical properties. Fullerenes became nanotech's first major new material. But what to do with them? Engineers turned their attention to finding some practical use for these interesting molecules.

While engineers thought about practical uses for fullerenes another discovery in search of an application was being made. In 1981 Gerd Karl Binnig and Heinrich Rohrer invented the scanning tunneling microscope or STM, which has a tiny tip so sensitive that it can in effect "feel" the surface of a single atom. It then sends information about the surface to a computer that reconstructs an image of the atomic surface on a display screen. If that weren't amazing enough, a little later, researchers discovered that the tip of the STM could actually move atoms around, and Donald Eigler and a team at IBM staged a dramatic demonstration of this new ability, spelling out "IBM". Researchers believed they had a tool, the atomic force microscope (AFM), that could build things atom-by-atom. But, like the discovery of fullerenes, it remained to be seen if anything useful could actually be built this way.

The development of tools such as AFMs coincided with the introduction of very powerful new computers and software that scientists could use to simulate and visualize chemical reactions or "build" virtual atoms and molecules. This was especially useful for scientists working with complex chemical molecules, particularly DNA.

Researchers recognized that the actions of DNA resembled some of the things nanotechnologists were now calling for-the use of molecules to construct other molecules, the self-replication of molecules, and the use of molecule-size mechanical devices. Perhaps DNA (or its cousin, RNA) could be modified to create the first nanomachines?

Geneticists had already found ways to use DNA taken from bacteria to make a nano-scale replicator used for scientific research. By modifying some of the chemical reactions that take place in natural DNA, genetic engineers had figured out a way to make copies of nearly any DNA molecule they wanted to study. But with the computers and tools available to them by the 1990s, they began using DNA or DNA-like molecules to do other things-like construct new chemicals or tiny machines. Many researchers began investigating ways to make proteins-the components from which DNA is made-that would perform useful tasks, such as interacting with other materials or living cells to create new materials or perhaps attack diseases. One of the first breakthroughs was Professor Nadrian Seeman's demonstration of a tiny "robot arm" made from modified DNA. While the arm could not yet really do anything useful, it did demonstrate the concept.

Meanwhile, electronics researchers approached nanotechnology from another direction. Since 1959, engineers had etched and coated silicon chips using a variety of processes to make integrated circuits (ICs). The transistors and other chip elements reached nano-scale in the late 1990s. They also used these same techniques to develop the first micromachines-microscopic devices with actual moving parts. Some of the early versions of these were simply intended to demonstrate the process without doing anything particularly useful, such as a tiny guitar with a string that could be plucked using an atomic force microscope. But in the late 1980s these began to be commercialized as machines-on-a-chip, or micro-electrco-mechanical systems (MEMs), which combine ICs and tiny mechanical elements. However useful MEMs are, most engineers feel that the techniques used to make ordinary ICs will never be refined enough to make true nanotechnologies. For that reason, engineers are now concentrating on discovering entirely new ways to make ICs, building them from the ground up rather than cutting and etching "bulk" silicon slices.

With the appearance of protein-based chemistry and other techniques in the 1990s, researchers began looking both for practical uses for nanotechnology and new ways to make nano-molecules or micromachines. A different but related problem was that of making nanomolecules in large numbers. A single nanomachine or nanocircuit for example, would not be able to do enough work to make a difference in the real world-thousands or millions might be needed. Engineers needed ways to turn out their nanomachines in huge numbers, and so they began looking for a way to make a nano-scale machine or molecule that would assemble other nano-scale machines or molecules. K. Eric Drexler called it a "self assembler," and scientists believe that it will be one of the keys to making certain kinds of nanotechnology useful and practical. To date, very few practical nanotechnologies and no self-assemblers have been used outside the laboratory.

THE PRESENT STATUS

Nanotechnology is a science in its infancy, but that doesn't mean it hasn't been put to use. What exactly has been accomplished in nanotechnology so far? In general, all of today's practical nanotechnologies are those using nano-size particles of various materials, or nanometer-size features on integrated circuits (ICs), rather than the complex molecular machines that engineers first envisioned. These current nanotechnologies are still made by "top down" methods (like those used in conventional chemistry and IC manufacturing), rather than the largely unproven "bottom up" techniques predicted by nanotechnology's boosters.

Many current nanotechnologies, for example, consist of the ever-shrinking transistors, interconnecting wires, and other features on digital ICs. As of 2005, some integrated circuits now have transistors that measure about 50 nanometers across-well inside the accepted size-based definition of nanotechnology. But chips are still made using advanced versions of the lithographic processes developed in the 1950s, which layer on materials and then carve away at them to form the electronic circuits. They are not, in other words, constructed molecule-by-molecule from the bottom up.

However, chip manufacturers point out that when working with extremely small circuit elements, the behavior of electrons changes, so entirely new principles are at work. Also, there is at least one new chip with a somewhat different claim to being "nanotechnological." This is IBM's "Millipede" memory chip, which draws its inspiration directly from the Atomic Force Microscope (AFM). Electronics manufacturers can also point to the latest generation of high-density computer hard drives, which have extremely thin coatings of just a few atoms' thickness applied to the surface of the disc by a process called chemical vapor deposition.

While such nanochips are beginning to appear in greater numbers, most of us more often encounter applications of nanotechnological materials that are made in "bulk" form and added to other products. By far the best known of these are the controversial "nanotechnology" trousers introduced by The Gap and Eddie Bauer stores in 2005. These were simply ordinary cotton pants, treated with nanoparticles of a new, stain-resistant chemical that attached itself to the cotton molecules.

Carbon nanotubes, which can now be made in large quantities at relatively low cost by companies like Hyperion Catalysis International Inc., are being incorporated into a wide range of other products. Because the fibers conduct electricity very well, Hyperion was able to mix them into plastic compounds, which auto makers can then mold into parts that conduct electricity. This is useful for preventing static electricity charges from building up on parts such as plastic fuel system components, where the static can eventually damage them or, in some cases, cause a spark.

Nanotubes mixed into plastics are very strong and light, and have been used to make car body components, tennis rackets, and other items. They have also been used to improve battery performance, and may some day be used in other technologies that traditionally used ordinary carbon or metals to conduct charges. Infineon Technologies in Germany, for example, has demonstrated the use of the tubes to connect components on microchips. In 2002 they showed how nanotubes could be used to replace ordinary metal wires allowing them to carry more current but taking up less space. That would result in computer chips that can pack more circuits into less space; one of the longstanding goals of chip designers.

One very useful new material is the semiconductor quantum dot. While not used in electronic circuits, quantum dots are nonetheless made from the same silicon used in computer chips. These tiny bits of material are coming into widespread use in experimental biology and, in a limited way, in medical diagnosis. The dots can be coated with certain chemicals, which are specially formulated so that they bind themselves to particular things-such as RNA, cell walls, or other types of molecules found in cells. One interesting application of this technology is its use in analyzing DNA material taken from the body. These DNA "scanners," first introduced commercially by Matsushita Corporation, combine integrated circuit technology and quantum dots to analyze genetic material much more rapidly than was possible before, and may lead to more rapid assessment of diseases.

A second use of coated quantum dots is injecting them into the body, where they circulate until they come in contact with whatever type of cell their coating is designed to attach itself to. Then when a powerful infrared light source is shone on the body, it penetrates the flesh, illuminates the massed quantum dots, and the reflections can be detected to provide a "live" picture of an organ, muscle, cancerous growth, or other internal part without the need for surgery. Unfortunately, not all of these quantum dots are suitable for injection into a living human body, and some are even poisonous, but bioengineers are working around that problem.

Even with these real-world applications, the current uses of nanotechnology (other than nano-size particles of various materials) remain very limited. In fact, several once-promising nanotechnology based systems introduced commercially in the 1990s did not meet with success, such as the nanotube-based Field Emission Displays proposed as competitors to other flat-panel information displays. However, researchers are rapidly making progress toward what some think of as true nanotechnologies-self-assembling, molecule size machines to perform all sorts of tasks (including manufacturing the nano-size materials made by other methods today). The nanotechnological future, we are told, is right around the corner.

NANOTECHNOLOGY IN COMING FUTURE

The future of nanotechnology is largely a question mark. Futurists say we are entering a new era, somewhat like the Industrial Revolution of the 18th and 19th centuries. That revolution changed nearly everything about the way people lived. But no one at that time could have predicted how those changes would unfold. Could we be on the brink of another very rapid period of profound technological and social change?

The nanotechnological revolution, if it occurs, will be just as unpredictable in the long-term, but scientists and engineers have laid out some pretty fantastic forecasts for the near-future. For example, some see great promise for the use of nanotubes in super-strong materials. Even though the plastic composites made today using relatively short nanotubes are not yet much stronger than earlier types of composites, long nanotubes are expected to be used for extraordinary applications like the proposed "space elevator." This system would replace rockets for the transport of payloads and people into earth orbit.

Another major area where nanotechnologists predict stunning changes is in medicine. Imagine a world where no one gets seriously ill, grows older, or even dies (until they want to). That is what the prophets of nanotechnology say is in store for the 21st century. Today's nanotechnologies used in medicine offer only modest benefits, such as the ability to target diseased or cancerous cells, making them easier to locate.

In the near future, engineers tell us, that will change. Tiny molecular machines, perhaps based on complex, branched molecules called "dendrimers" will be injected into the body not only to locate cancers but also to find and repair cells damaged by disease or aging. Livers and hearts damaged by natural wear-and-tear, inherited diseases, poor nutrition, or alcoholism will be fixed or even replaced. Genetically based ailments such as Alzheimer's will be cured by replacing the faulty genes.

Some futurists have predicted that the most profound changes will be the result of the introduction of molecular assembly "factories," perahps even small small enough to fit on a desktop. These would, some say, make it possible for virtually anyone to design and build virtually anything, using nanorobots or perhaps a new technology called "nanoink," created by NanoInk founder Chad Mirkin. Nanotechnology researchers like K. Eric Drexler and Ralph Merkle say that this could be done by imitating and improving upon the "manufacturing" that DNA accomplishes inside the body. In 1999, researchers such as Nadrian Seeman at New York University demonstrated the principle of using modified DNA molecules to build a tinymachine, and somewhat later Nanoink founder Chad Mirkin had demonstrated building up nanostructures by depositing layers of materials on a substrate.

These and other experiments have also led researchers to believe that they will eventually be able to assemble circuits atom-by-atom in order to create the next generation of computer chips. The circuits on these chips will be much smaller than what is currently possible, and will enable the building of much more powerful computers. What difference will that make? Some, like engineer Raymond Kurzweil, think that computers will have personalities and be as smart as humans within 20 years. We may even be able to "download" our own personalities into computers, to become virtual humans. With nearly unlimited computing power, programmers are sure they could create software that completely blows away anything possible today.

Not surprising, these amazing predictions have inspired fear as well as wonder. Environmentalists and others point out that nanotechnology may bring with it unexpected dangers. The nanomaterials being made today, like fullerenes, are often in the form of extremely small particles. Even when these particles are made from common materials like carbon, they may interact with the human body or the environment in ways that are unlike those of natural particles of the same materials. Some say that allowing nanoparticles to be included in products ingested or applied to the body may pose health risks for consumers. Others predict that nanotechnology may get out of control, causing a huge man-made disaster. Eric Drexler and others, such as computer engineer Bill Joy of Sun Microsystems, warned in 2000 that self-replicating machines might run amok if they escape into the environment, competing with natural bacteria, plants, and people for natural resources. Then, in 2002 the public's awareness of nanotechnology-the bad side of nanotechnology-was greatly expanded when author Michael Crichton published his best-selling novel Prey, about tiny, self-duplicating nanorobots that band together to try to take over the world.

Whether public fears are founded in fact, it is true that the future of nanotechnology has inspired as much caution as optimism. Recently, in response to public outcry, researchers such as Dr. Vicky Colvin of Rice University have begun evaluating the risks and rewards of current nanotechnologies. Colvin and other engineers believe that, with wisdom, they can bring the wonders of nanotechnology into being while avoiding the pitfalls.

PAVING THE WAY FOR SMALL MACHINES

Disposable satellite transmitters, inexpensive medical testing equipment and sensors for automatically tracking inventory or traffic patterns will become possible over the next 10 years through developments in nanotechnology, speakers at the Nanotech 2003 conference said Monday.

Nanotechnology-the science of making devices with features measuring less than 100 nanometers (or one-ten-millionth of a meter)-will let companies make smaller and cheaper products than ever before, which in turn will lead to new markets, according to Albert Pisano, a professor of engineering and computer science at the University of California at Berkeley.

Communications, for instance, could be greatly affected, Pisano noted. Right now, radios on satellites have to be hardened against radiation, an expensive process. At Stanford University, researchers have shown how small, unhardened radios can transmit and receive messages while in orbit. Although these smaller radios are subject to harm from radiation, they cost far less, so many could be mounted on a spacecraft to compensate for burnouts.

A market for tiny transmitters could also emerge within five years, out of a need to better monitor power lines, freeways and bridges for potential failures.

The movement will also mark a major turning point in mechanical engineering because most of these tiny systems will deal directly with the physical world, like pumps do, rather than getting information from electrical impulses, like microprocessors.

"Most people forget that a radio is two-thirds filter and that filters are mechanical," Pisano said. "Sensors and computing and communications can all be heavily miniaturized."

Batteries, too, could improve through breakthroughs in mechanical engineering. Currently, a kilogram of lithium ion, the same material used in notebook batteries, provides about 400 watt-hours of energy. A kilogram of methane provides 15,000 watt-hours of energy. The difference is that lithium ion is electrically charged, while methane provides thermal energy-that is, heat.

"Can you use thermal energy on such a small scale?" Pisano asked. Apparently so. Some researchers have already developed rotary engines that measure only microns across.

Of course, getting to miniature nirvana won't be easy. In the early '90s, futurists predicted a booming market for MEMS (microelectromechanical systems) similar to the devices described above. Difficulties in mass manufacturing, packaging and other problems kept the market from taking flight.

Pisano, though, noted that new technologies take time. About 15 years passed between the development of the transistor in the late '40s to the true development of the electronics industry. Several challenges need to be overcome, but the progress with nanotechnology makes MEMS more promising.

The first must-have applications for the types of MEMS products Pisano described will appear in about five years, he predicted, because the demand exists. Gillette, for instance, has spent heavily in the last few months on diminutive radio-frequency transmitters.

A number of universities and private companies will be presenting papers on MEMS over the five-day conference here. Some of the major themes include microfluidics, or small machines that can test biological samples, and films or surfaces with sensory capabilities.

Other companies are presenting papers on progress in using chains of molecules to make processors or memory devices. Molecular computing will likely take much longer to develop, but early results are showing promise.

Molecular chips differ substantially from today's silicon processors, noted Phaedon Avouris, manager of nanoscience and nanotechnology at IBM's Watson Research Center. Carbon nanotubes exposed to the air, for instance, carry a positive charge. When hermetically sealed off, they carry a negative charge.

SMALL TRUTH OF NANOTECH

The drop-off was stomach-churning. In 2000, venture capitalists poured $100 billion into startups. Last year, they couldn't even reach $40 billion. So forgive them for latching onto nanotech as the uptrend. Headlines like "NANOTECHNOLOGY WINS OVER MAINSTREAM VENTURE CAPITALISTS" and "THE NEXT BIG THING IS VERY SMALL" are getting hard to avoid. Nano conferences are weekly events, crowded with VCs amped up about self-assembling machines and nanobots in your bloodstream. But what's really going on? While venture capitalists are happy to hype nanotech, they aren't exactly rushing in where angel investors fear to tread. Of the $5.6 billion put up by private capital sources in the first three months of this year, only $42 million went to very small tech, according to Steve Glapa, president of In Realis, a market consulting firm in Milpitas, California. Even putting all of these investments under the "nanotech" umbrella is a stretch. Advanced chemistry and materials science are more like it.

The reality of nanotechnology has less to do with self-assembling Ferraris than with powders and filters and goop. It's about taking a material like zinc oxide and grinding it down to grains the size of a few dozen nanometers (great for sunscreen). Or punching nano-size holes in screens so hydrogen atoms can slip through but water molecules can't (handy for fuel cells). Or laying down slurries a few atoms thick (good for hard drives).

It's not a new industry so much as the outcome of decades of research. "People have been doing this sort of thing for a long time," says Glapa. "We just didn't always attach the nano prefix to it."

So why all the fuss from VCs now? Why the conferences, the speeches, the sound bites? Because it's great business. Attaching nano to a company name can instantly multiply the value. Or at least that's the idea. "VCs invest in companies where they see an exit strategy," says Glenn Fishbine, author of The Investor's Guide to Nanotechnology and Micromachines. "The more they can do to make nanotechnology attractive now, the better off they'll be when it's time to sell."

Unlike past technology booms, it's unclear whether VCs will ever have much of a role to play in the nanotech realm. Industrial labs at IBM, Hewlett-Packard, and 3M have invested hundreds of millions of dollars in scaling materials down to the atomic level, a gap that VCs are unlikely to close. "They are playing in exactly the same market," says Tom Theis, IBM's director of materials research. "They have no advantage over what big companies have looked at and rejected."

The structure of VC investing- seed money, ramp-up, IPO, exit- won't work easily with nanotech. It's an area where $800,000 government grants are readily available, often enough for a research scientist to get to a working prototype. The next step, bringing a product to market, will often take longer than VCs are willing to wait. So researchers with promising technologies are likely to bypass VCs altogether and sell directly to an established industrial player. That's the route Nano-Tex took when the Emeryville, California-based company developed a nanoscale technology to protect cotton from stains, then sold it to Burlington. "If someone has a really good memory technology, they won't go to a VC, they'll just go to IBM," says Fishbine. "Then they don't have to worry about marketing and manufacturing and distribution; they'll just take the royalty payments and buy land on Kauai."

Not so fast, says Steve Jurvetson of Valley VC firm Draper Fisher Jurvetson. Before startups can partner with big industrial players or license intellectual property to them, they need backing from venture capitalists. At the same time, he adds, VCs need to relearn patience. During the Internet craze, the moneybags on Sand Hill Road got used to going from concept to IPO in as little as 18 months. Post-bust, VCs are returning to the standard three- to seven-year wait for a return. "I think we're on the long end of that for nanotech," says Jurvetson, adding that a dozen years isn't out of the question. "If the opportunity is big enough, if it's really going to change the world in a significant way, then it's worth the wait."

That thinking- and time horizon- may work for today's nano of chemistry and goop. But the VCs are luring investors with a grander vision: the idea that someday we might manipulate matter in fantastic ways, that we could build complex structures from the atomic level- much the way living things self-assemble from info stored in their DNA. Of course, that's all speculative stuff, and even nano evangelists like Richard Smalley say we're decades from realizing that dream, well beyond even Jurvetson's expanded VC timeline.

So don't believe the hype. The next great nanotech triumph is not going to be about visionary investment mavericks who see the truth long before the dull, slow multinationals. It's going to be about someone slogging away in a basement lab for 15 years. This time around, the deep-pocketed corporations will lead the charge. To do this story justice, we're going to need a brand-new script. Less about the money, more about the science. Less sizzle, more steak.

7

THE APPLIED FIELD OF NANOTECHNOLOGY

Nanotechnology is a field of applied science and technology covering a broad range of topics. The main unifying theme is the control of matter on a scale smaller than 1 micrometer, as well as the fabrication of devices on this same length scale. It is a highly multidisciplinary field, drawing from fields such as colloidal science, device physics, and supramolecular chemistry. Much speculation exists as to what new science and technology might result from these lines of research. Some view nanotechnology as a marketing term that describes pre-existing lines of research applied to the sub-micron size scale.

Despite the apparent simplicity of this definition, nanotechnology actually encompasses diverse lines of inquiry. Nanotechnology cuts across many disciplines, including colloidal science, chemistry, applied physics, materials science, and even mechanical and electrical engineering. It could variously be seen as an extension of existing sciences into the nanoscale, or as a recasting of existing sciences using a newer, more modern term. Two main approaches are used in nanotechnology: one is a "bottom-up" approach where materials and devices are built from molecular components which assemble themselves chemically using principles of molecular recognition; the other being a "top-down" approach where nano-objects are constructed from larger entities without atomic-level control.

The impetus for nanotechnology has stemmed from a renewed interest in colloidal science, coupled with a new generation of analytical tools such as the atomic force microscope (AFM) and the scanning tunneling microscope (STM). Combined with refined processes such as electron beam lithography and molecular beam epitaxy, these instruments allow the deliberate manipulation of nanostructures, and in turn led to the observation of novel phenomena. The manufacture of polymers based on molecular structure, or the design of computer chip layouts based on surface science are examples of nanotechnology in modern use. Despite the great promise of numerous nanotechnologies such as quantum dots and nanotubes, real applications that have moved out of the lab and into the marketplace have mainly utilized the advantages of colloidal nanoparticles in bulk form, such as suntan lotion, cosmetics, protective coatings, and stain resistant clothing.

GROWTH AND ORIGIN

The first mention of some of the distinguishing concepts in nanotechnology (but predating use of that name) was in "There's Plenty of Room at the Bottom," a talk given by physicist Richard Feynman at an American Physical Society meeting at Caltech on December 29, 1959. Feynman described a process by which the ability to manipulate individual atoms and molecules might be developed, using one set of precise tools to build and operate another proportionally smaller set, so on down to the needed scale.

In the course of this, he noted, scaling issues would arise from the changing magnitude of various physical phenomena: gravity would become less important, surface tension and Van der Waals attraction would become more important, etc. This basic idea appears feasible, and exponential assembly enhances it with parallelism to produce a useful quantity of end products.

The term "nanotechnology" was defined by Tokyo Science University Professor Norio Taniguchi in a 1974 paper

Nanotechnology and nanoscience got started in the early 1980s with two major developments; the birth of cluster science and the invention of the scanning tunneling microscope (STM). This development led to the discovery of fullerenes in 1986 and carbon nanotubes a few years later. In another development, the synthesis and properties of semiconductor nanocrystals was studied. This led to a fast increasing number of metal oxide nanoparticles of quantum dots.

BASIC CONCEPTS

One nanometer (nm) is one billionth, or 10-9 of a meter. For comparison, typical carbon-carbon bond lengths, or the spacing between these atoms in a molecule, are in the range .12-.15 nm, and a DNA double-helix has a diameter around 2 nm. On the other hand, the smallest cellular life-forms, the bacteria of the genus Mycoplasma, are around 200 nm in length.

A Materials Perspective

A unique aspect of nanotechnology is the vastly increased ratio of surface area to volume present in many nanoscale materials which opens new possibilities in surface-based science, such as catalysis. A number of physical phenomena become noticeably pronounced as the size of the system decreases. These include statistical mechanical effects, as well as quantum mechanical effects, for example the "quantum size effect" where the electronic properties of solids are altered with great reductions in particle size. This effect does not come into play by going from macro to micro dimensions. However, it becomes dominant when the nanometer size range is reached. Additionally, a number of physical properties change when compared to macroscopic systems. One example is the increase in surface area to volume of materials. This catalytic activity also opens potential risks in their interaction with biomaterials.

Materials reduced to the nanoscale can suddenly show very different properties compared to what they exhibit on a macroscale, enabling unique applications. For instance, opaque substances become transparent (copper); inert materials become catalysts (platinum); stable materials turn combustible (aluminum); solids turn into liquids at room temperature (gold); insulators become conductors (silicon). Materials such as gold, which is chemically inert at normal scales, can serve as a potent chemical catalyst at nanoscales. Much of the fascination with nanotechnology stems from these unique quantum and surface phenomena that matter exhibits at the nanoscale.

A Molecular Perspective

Modern synthetic chemistry has reached the point where it is possible to prepare small molecules to almost any structure. These methods are used today to produce a wide variety of useful chemicals such as pharmaceuticals or commercial polymers. This ability raises the question of extending this kind of control to the next-larger level, seeking methods to assemble these single molecules into supramolecular assemblies consisting of many molecules arranged in a well defined manner.

These approaches utilize the concepts of molecular self-assembly and/or supramolecular chemistry to automatically arrange themselves into some useful conformation through a bottom-up approach. The concept of molecular recognition is especially important: molecules can be designed so that a specific conformation or arrangement is favored due to non-covalent intermolecular forces. The Watson-Crick basepairing rules are a direct result of this, as is the specificity of an enzyme being targeted to a single substrate, or the specific folding of the protein itself. Thus, two or more components can be designed to be complementary and mutually attractive so that they make a more complex and useful whole.

Such bottom-up approaches should, broadly speaking, be able to produce devices in parallel and much cheaper than top-down methods, but could potentially be overwhelmed as the size and complexity of the desired assembly increases. Most useful structures require complex and thermodynamically unlikely arrangements of atoms. The basic laws of probability and entropy make it difficult to self-assemble molecules in useful configurations. Nevertheless, there are many examples of self-assembly based on molecular recognition in biology, most notably Watson-Crick basepairing and enzyme-substrate interactions. The challenge for nanotechnology is whether these principles can be used to engineer novel constructs in addition to natural ones.

A Long-term View

Molecular nanotechnology, sometimes called molecular manufacturing, is a term given to the concept of engineered nanosystems (nanoscale machines) operating on the molecular scale. It is especially associated with the concept of a molecular assembler, a machine that can produce a desired structure or device atom-by-atom using the principles of mechanosynthesis. Manufacturing in the context of productive nanosystems is not related to, and should be clearly distinguished from, the conventional technologies used to manufacture nanomaterials such as carbon nanotubes and nanoparticles. It should also be noted that, while Drexler's vision is the source of the concepts of nanorobots and gray goo in popular culture, these belong in the realm of science fiction are not possible with any technology currently foreseeable.

When the term "nanotechnology" was independently coined and popularized by Eric Drexler (who at the time was unaware of an earlier usage by Norio Taniguchi) it referred to a future manufacturing technology based on molecular machine systems. The premise was that molecular-scale biological analogies of traditional machine components demonstrated that molecular machines were possible: by the countless examples found in biology, it is known that billions of years of evolutionary feedback can produce sophisticated, stochastically optimized biological machines. It is hoped that developments in nanotechnology will make possible their construction by some other means, perhaps using biomimetic principles.

However, Drexler and other researchers have proposed that advanced nanotechnology, although perhaps initially implomented by biomimetic means, ultimately could be based on mechanical engineering principles, namely, a manufacturing technology based on the mechanical functionality of these components (such as gears, bearings, motors, and structural members) that would enable programmable, positional assembly to atomic specification (PNAS-1981). The physics and engineering performance of exemplar designs were analyzed in Drexler's book Nanosystems. But Drexler's analysis is very qualitative and does not address very pressing issues, such as the "fat fingers" and "Sticky fingers" problems. In general it is not possible to assemble devices on the atomic scale, as all one has to position atoms are other atoms of comparable size and stickyness. Drexler also glosses over the numerical challenges to nano-construction--assembling just one mole's worth of nanodevices at the rate of a billion atoms per second would take 19 million years.

Another view, put forth by Carlo Montemagno, is that future nanosystems will be hybrids of silicon technology and biological molecular machines. Yet another view, put forward by the late Richard Smalley, is that mechanosynthesis is impossible due to the difficulties in mechanically manipulating individual molecules. This lead to an exchange of letters in the ACS publication Chemical & Engineering News in 2003.

Though biology clearly demonstrates that molecular machine systems are possible, non-biological molecular machines are today only in their infancy. Leaders in research on non-biological molecular machines are Dr. Alex Zettl and his colleagues at Lawrence Berkeley Laboratories and UC Berkeley. They have constructed at least three distinct molecular devices whose motion is controlled from the desktop with changing voltage: a nanotube nanomotor, a molecular actuator, and a nanoelectromechanical relaxation oscillator. An experiment indicating that positional molecular assembly is possible was performed by Ho and Lee at Cornell University in 1999. They used a scanning tunneling microscope to move an individual carbon monoxide molecule (CO) to an individual iron atom (Fe) sitting on a flat silver crystal, and chemically bound the CO to the Fe by applying a voltage.

RECENT RESEARCHES IN THE FIELD

As nanotechnology is a very broad term, there are many disparate but sometimes overlapping subfields that could fall under its umbrella. The following avenues of research could be considered subfields of nanotechnology. Note that these categories are fairly nebulous and a single subfield may overlap many of them, especially as the field of nanotechnology continues to mature.

Nanomaterials

This includes subfields which develop or study materials having unique properties arising from their nanoscale dimensions.

- o Colloid science has given rise to many materials which may be useful in nanotechnology, such as carbon nanotubes and other fullerenes, and various nanoparticles and nanorods.
- o Nanoscale materials can also be used for bulk applications; most present commercial applications of nanotechnology are of this flavor.
- o Headway has been made in using these materials for medical applications; see Nanomedicine.

Bottom-up Approaches

These seek to arrange smaller components into more complex assemblies.

- o DNA Nanotechnology utilizes the specificity of Watson-Crick basepairing to construct well-defined structures out of DNA and other nucleic acids.
- o More generally, molecular self-assembly seeks to use concepts of supramolecular chemistry, and molecular recognition in particular, to cause single-molecule components to automatically arrange themselves into some useful conformation.

Top-down Approaches

These seek to create smaller devices by using larger ones to direct their assembly.

- Many technologies descended from conventional solid-state silicon methods for fabricating microprocessors are now capable of creating features smaller than 100 nm, falling under the definition of nanotechnology. Giant magnetoresistance-based hard drives already on the market fit this description, as do atomic layer deposition (ALD) techniques.
- Solid-state techniques can also be used to create devices known as nanoelectromechanical systems or NEMS, which are related to microelectromechanical systems or MEMS.
- Atomic force microscope tips can be used as a nanoscale "write head" to deposit a chemical on a surface in a desired pattern in a process called dip pen nanolithography. This fits into the larger subfield of nanolithography.

Functional Approaches

These seek to develop components of a desired functionality without regard to how they might be assembled.

- Molecular electronics seeks to develop molecules with useful electronic properties. These could then be used as single-molecule components in a nanoelectronic device.
- Synthetic chemical methods can also be used to create synthetic molecular motors, such as in a so-called nanocar.

Speculative

These subfields seek to anticipate what inventions nanotechnology might yield, or attempt to propose an agenda along which inquiry might progress. These often take a big-picture view of nanotechnology, with more emphasis on its societal implications than the details of how such inventions could actually be created.

- o Molecular nanotechnology is a proposed approach which involves manipulating single molecules in finely controlled, deterministic ways. This is more theoretical than the other subfields and is beyond current capabilities.
- o Nanorobotics centers on self-sufficient machines of some functionality operating at the nanoscale.
- o Programmable matter based on artificial atoms seeks to design materials whose properties can be easily and reversibly externally controlled.
- o Due to the popularity and media exposure of the term nanotechnology, the words picotechnology and femtotechnology have been coined in analogy to it, although these are only used rarely and informally.

THE TOOLS AND TECHNIQUES

Nanotechnological techniques include those used for fabrication of nanowires, those used in semiconductor fabrication such as deep ultraviolet lithography, electron beam lithography, focused ion beam machining, nanoimprint lithography, atomic layer deposition, and molecular vapor deposition, and further including molecular self-assembly techniques such as those employing di-block copolymers. However, all of these techniques preceded the nanotech era, and are extensions in the development of scientific advancements rather than techniques which were devised with the sole purpose of creating nanotechnology and which were results of nanotechnology research.

Nanoscience and nanotechnology only became possible in the 1910s with the development of the first tools to measure and make nanostructures. But the actual development started with the discovery of electrons and neutrons which showed scientists that matter can really exist on a much smaller scale than what we normally think of as small, and/or what they thought was possible at the time. It was at this time when curiosity for nanostructures had originated.

The atomic force microscope (AFM) and the Scanning Tunneling Microscope (STM) are two early versions of scanning probes that launched nanotechnology. There are other types of scanning probe microscopy, all flowing from the ideas of the scanning confocal microscope developed by Marvin Minsky in 1961 and the scanning acoustic microscope (SAM) developed by Calvin Quate and coworkers in the 1970s, that made it possible to see structures at the nanoscale. The tip of a scanning probe can also be used to manipulate nanostructures (a process called positional assembly). However, this is a very slow process. This led to the development of various techniques of nanolithography such as dip pen nanolithography, electron beam lithography or nanoimprint lithography. Lithography is a top-down fabrication technique where a bulk material is reduced in size to nanoscale pattern.

The top-down approach anticipates nanodevices that must be built piece by piece in stages, much as manufactured items are currently made. Scanning probe microscopy is an important technique both for characterization and synthesis of nanomaterials. Atomic force microscopes and scanning tunneling microscopes can be used to look at surfaces and to move atoms around.

By designing different tips for these microscopes, they can be used for carving out structures on surfaces and to help guide self-assembling structures. Atoms can be moved around on a surface with scanning probe microscopy techniques, but it is cumbersome, expensive and very time-consuming. For these reasons, it is not feasible to construct nanoscaled devices atom by atom. Assembling a billion transistor microchips at the rate of about one transistor an hour is inefficient.

In contrast, bottom-up techniques build or grow larger structures atom by atom or molecule by molecule. These techniques include chemical synthesis, self-assembly and positional assembly. Another variation of the bottom-up approach is molecular beam epitaxy or MBE. Researchers at Bell Telephone Laboratories like John R. Arthur. Alfred Y. Cho, and Art C. Gossard developed and implemented MBE as a research tool in the late 1960s and 1970s. Samples made by MBE were key to to the discovery of the fractional quantum Hall effect for which the 1998 Nobel Prize in Physics was awarded. MBE allows scientists to lay down atomically-precise layers of atoms and, in the process, build up complex structures. Important for research on semiconductors, MBE is also widely used to make samples and devices for the newly emerging field of spintronics.

Newer techniques such as Dual Polarisation Interferometry are enabling scientists to measure quantitatively the molecular interactions that take place at the nano-scale.

APPLICATIONS

Although there has been much hype about the potential applications of nanotechnology, most current commercialized applications are limited to the use of nanomaterials in bulk, for example the use of titanium dioxide nanoparticles to make transparent sunscreen. Further applications which require actual manipulation or arrangement of nanoscale components await further research.Though technologies currently branded with the term 'nano' are sometimes little related to and fall far short of the most ambitious and transformative technological goals of the sort in molecular manufacturing proposals, the term still connotes such ideas. Thus there may be a danger that a "nano bubble" will form, or is forming already, from the use of the term by scientists and entrepreneurs to garner funding, regardless of interest in the transformative possibilities of more ambitious and far-sighted work. So far about $400 million has been invested in nanotechnology, with rather meager results.

THE BREAK JUNCTION

A break junction is an electrical junction between two wires formed by pulling the wires apart to produce electrodes separated by a few atomic distances. In this technique a metal wire is bent or pulled, often using a piezoelectric crystal to apply the necessary force. The bending or pulling causes the metal wire to break in a controlled manner since piezoelectric elongation can be controlled to a precision of angstroms or less (such crystals are used for motion control in scanning tunneling microscopy). As the wire breaks, the separation between the electrodes can be indirectly controlled by monitoring the electrical current through the junction.

A typical conductance versus time trace during the breaking process (conductance is simply current divided by applied voltage bias) shows two regimes. First is a regime where the break junction comprises a quantum point contact. In this regime conductance decreases in steps equal to the conductance quantum GQ = 2e2/h which is expressed through the electron charge e and Planck's constant h. The conductance quantum has a value of 7.74 × 10-5 Siemens, corresponding to a resistance increase of roughly 12.9 KO. These step decreases are interpreted as the result of a decrease, as the electrodes are pulled apart, in the number of single-atom-wide metal strands bridging between the two electrodes, each strand having a conductance equal to the quantum of conductance. As the wire is pulled, the neck becomes thinner with fewer atomic strands in it. Each time the neck reconfigures, which happens abruptly, a step-like decrease of the conductance can be observed. This picture inferred from the current measurement has been confirmed by "in-situ" TEM imaging of the breaking process combined with current measurement. In a second regime, when the wire is pulled further apart, the conductance collapses to values less than the quantum of conductance. This is the tunneling regime where electrons tunnel through vacuum between the electrodes. Digging into the literature on break junctions and quantum point contacts reveals that the above conceptual description is somewhat oversimplified, but the description is a good first approach to understanding the topic.

Use

This method has been developed to study the conductance of few-atom constrictions of varied metals. The conductance of these constrictions has been compared with theoretical predictions for both the stability and the conductance of possible few-atom configurations. More recently it has been used to study molecules which are inserted in the junction in the liquid phase and binds to them (dithiols) or in the gas phase. This method has several advantages. It is clean, since the junctions can be made in a controlled atmosphere (high vacuum). It is fast and thus enables many independent measurements to be done in a few hours. It is then possible to study the statistical occurrence of a particular type of contact, and build conductance histograms. Lately this method has enabled the more accurate determination of the conduction of a single molecule. The disadvantage of this technique is that it is a two-terminal technique (that is, it uses only two wires and can be considered an electrical diode), whereas complete determination of electronic properties requires using a three-terminal configuration similar to the source, drain and gate of an MOS transistor.

MICROMACHINERY

Micromachines are mechanical objects that are fabricated in the same general manner as integrated circuits. They are generally considered to be between 100 nanometres to 100 micrometres in size, though that is debatable. The applications of micromachines include accelerometers that detect when a car has hit an object and trigger an airbag. Complex systems of gears and levers are another application.

The fabrication of these devices is usually done by one or both of two techniques: surface micromachining and bulk micromachining.

Most micromachines act as transducers; in other words, they are either sensors or actuators.

Sensors convert information from the environment into interpretable electrical signals. One example of a micromachine sensor is a resonant chemical sensor. A lightly damped mechanical object vibrates much more at one frequency than any other, and this frequency is called its resonant frequency. A chemical sensor is coated with a special polymer that attracts certain molecules, such as anthrax, and when those molecules attach to the sensor, its mass increases. The increased mass alters the resonant frequency of the mechanical object, which is detected with circuitry.

Actuators convert electrical signals and energy into motion of some kind. The three most common types of actuators are electrostatic, thermal, and magnetic. Electrostatic actuators use the force of electrostatic energy to move objects. Two mechanical elements, one that is stationary (the stator) and one that is movable (the rotor) have two different voltages applied to them, which creates an electric field. The field competes with a restoring force on the rotor (usually a spring force produced by the bending or stretching of the rotor) to move the rotor. The greater the electric field, the farther the rotor will move. Thermal actuators use the force of thermal expansion to move objects. When a material is heated, it expands and amount depending on material properties. Two objects can be connected in such a way that one object is heated more than the other and expands more, and this imbalance creates motion. The direction of motion depends on the connection between the objects. This is seen in a "heatuator", which is a U-shaped beam with one wide arm and one narrow arm. When a current is passed through the object, heat is created. The narrow arm is heated more than the wide arm due to the fact that they have the same current density. Since the two arms are connected at the top, the stretching hot arm pushes in the direction of the cold arm. Magnetic actuators used fabricated magnetic layers to create forces.

NANO-ABACUS

The nano-abacus is a nano-sized abacus developed by IBM scientist. Stable rows made up of ten molecules act as the railings of the abacus. The beads are made up of fullerene and are pushed around by a scanning tunneling microscope tip. The nano-abacus has the potential to be used in a variety of nanotechnological inventions such as the nano-computer.

NANOMOTOR

A nanomotor is a molecular device capable of converting energy into movement and forces on the order of the piconewtons. A proposed branch of research is the integration of molecular motor proteins found in living cells into molecular motors implanted in artificial devices. Such a motor protein would be able to move a "cargo" within that device, similarly to how kinesin moves various molecules along tracks of microtubules inside cells.

Starting and stopping the movement of such motor proteins would involve caging the ATP in molecular structures sensitive to UV light, pulses of UV illumination would thus provide pulses of movement.

Nanomotors have also be made using synthetic materials and chemical methods, as described in the following section.

NANOTUBE AND NANOMOTOR

Researchers at University of California, Berkeley, have developed rotational bearings based upon multiwall carbon nanotubes. By attaching a gold plate (with dimensions of order 100nm) to the outer shell of a suspended multiwall carbon nanotube (like nested carbon cylinders), they are able to electrostatically rotate the outer shell relative to the inner core. These bearings are very robust; Devices have been oscillated thousands of times with no indication of wear. The work was done in situ in an SEM. These nanoelectromechanical systems (NEMS) are the next step in miniaturization that may find their way into commercial aspects in the future.

Notice: The thin vertical string seen in the middle, is the nanotube to which the rotor is attached. When the outer tube is sheared, the rotor is able to spin freely on the nanotube bearing.

The process and technology can be seen in this render.

- o Physicists build world's smallest motor using nanotubes and etched silicon

- o Research Project
- o Carbon nanotube
- o Electrostatic motor

Nanopore

A nanopore is a small pore in an electrically insulating membrane, that can be used as a single-molecule detector. It can be a biological protein channel in a lipid bilayer or a pore in a solid-state membrane. The detection principle is based on monitoring the ionic current of an electrolyte solution passing through the nanopore as a voltage is applied across the membrane. When the nanopore is of molecular dimensions, passage of molecules (e.g., DNA) cause interruptions of the "open" current level, leading to a "translocation event" pulse. The passage of single-stranded DNA molecules through the membrane-embedded alpha-hemolysin channel (1.5 nm diameter), for example, causes a ~90% blockage of the current (measured at 1 M KCl solution). The observation that a passing strand of DNA containing different bases results in different blocking levels has led to the nanopore sequencing hypothesis. Such sequencing, if successful, could revolutionize the field of genomics, as sequencing could be carried out in a matter of seconds.Apart from rapid DNA sequencing, other applications include separation of single stranded and double stranded DNA in solution, and the determination of length of polymers. At this stage, nanopores are making contributions to the understanding of polymer biophysics, as well as to single-molecule analysis of DNA-protein interactions.

Solid-state nanopores are generally made in silicon compound membranes, one of the most common being Si3N4. Solid-state nanopores can be manufactured with several techniques including ion-beam sculpting and electron beams.

THE NANOCIRCUITS

Nanocircuits are electrical circuits on the scale of nanometers. One nanometer is equal to 10-9 meters or a row of 10 hydrogen atoms. With circuits becoming smaller, they are able to fit more on a computer chip. Thus, they will be able to perform more complex functions using less power and at a faster speed. Nanocircuits are organized into three different parts: transistors, interconnections, and architecture, all dealt within the nano scale.

Method of Production

One of the most fundamental concepts to understanding nanocircuits is the formulation of Moore's Law. This concept arose when Intel co-founder Gordon Moore became interested in the cost of transistors and trying to fit more onto one chip. It relates that the number of transistors that can be fabricated on a silicon integrated circuit-and therefore the computing speed of such a circuit-is doubling every 18 to 24 months. (1) The more transistors one can fit on a circuit, the faster the computer will be. This is why scientists and engineers are working together to produce these nanocircuits so millions and perhaps even billions of transistors will be able to fit onto a chip. Despite how good this may sound, there are many problems that arise when so many transistors are packed together. With circuits being so tiny, they tend to have more problems than larger circuits, more particularly many defects. Nanoscale circuits are more sensitive to temperature changes, cosmic rays and electromagnetic interference than today's circuits. (2) As they pack more transistors onto a chip, phenomena such as stray signals on the chip, the need to dissipate the heat from so many closely packed devices, and the difficulty of creating the devices in the first place will halt or severely slow progress.

(3) Many believe the market for nanocircuits will reach equilibrium around 2015. At this time they believe the cost of a fabrication facility may be as much as $200 billion. There will be a time when the cost of making circuits even smaller will be too much, and the speed of computers will reach a maximum. For this reason, many scientists believe that Moore's Law will not hold forever and will soon reach a peak.

In producing these nanocircuits, there are many aspects involved. The first part of their organization begins with transistors. As of right now, most electronics are using silicon-based transistors. Transistors are an integral part of circuits as they control the flow of electricity and transform weak electrical signals to strong ones. They also control electric current as they can turn it on off, or even amplify signals. Circuits now use silicon as a transistor because it can easily be switched between conducting and nonconducting states. However, in nanoelectronics, transistors might be organic molecules or nanoscale inorganic structures. Semiconductors, which are part of transistors, are also being made of organic molecules in the nano state.

The second aspect of nanocircuit organization is interconnection. This involves logical and mathematical operations and the wires linking the transistors together that make this possible. In nanocircuits, nanotubes and other wires as narrow as one nanometer are used to link transistors together. Nanowires have been made from carbon nanotubes for a few years. Until a few years ago, transistors and nanowires were put together to produce the circuit. However, scientists have been able to produce a nanowire with transistors in it. In 2004, Harvard University nanotech pioneer Charles Lieber and his team have made a nanowire-10,000 times thinner than a sheet of paper-that contains a string of transistors. Essentially, transistors and nanowires are already pre-wired so as to eliminate the difficult task of trying to connect transistors together with nanowires. The last part of nanocircuit organization is architecture. This has been explained as the overall way the transistors are interconnected, so that the circuit can plug into a computer or other system and operate independently of the lower-level details. With nanocircuits being so small, they are destined for error and defects. Scientists have devised a way to get around this. Their architecture combines circuits that have redundant logic gates and interconnections with the ability to reconfigure structures at several levels on a chip. The redundancy lets the circuit identify problems and reconfigure itself so the circuit can avoid more problems. It also allows for errors within the logic gate and still have it work properly without giving a wrong result.

APPLICATIONS AND THE BREAKTHROUGHS

Scientists in India have recently developed the world's smallest transistor which will be used for nanocircuits. The transistor is made entirely from carbon nanotubes. Nanotubes are rolled up sheets of carbon atoms and are more than a thousand times thinner than human hair. Normally circuits use silicon-based transistors, but these will soon replace those. The transistor has two different branches that meet at a single point, hence giving it a Y shape. Current can flow throughout both branches and is controlled by a third branch that turns the voltage on or off. This new breakthrough can now allow for nanocircuits to hold completely to their name as they can be made entirely from nanotubes. Before this discovery, logic circuits used nanotubes, but needed metal gates to be able to control the flow of electrical current.

Arguably the biggest potential application of nanocircuits deals with computers and electronics.

Scientists and engineers are always looking to make computers faster. Some think in the nearer term, we could see hybrids of micro and nano: silicon with a nano core-perhaps a high-density computer memory that retains its contents forever. Unlike conventional circuit design, which proceeds from blueprint to photographic pattern to chip, nanocircuit design will probably begin with the chip-a haphazard jumble of as many as 1024 components and wires, not all of which will even work-and gradually sculpt it into a useful device. Instead of taking the traditional top-down approach, the bottom-up approach will probably soon have to be adopted because of the sheer size of these nanocircuits. Not everything in the circuit will probably work because at the nano level, nanocircuits will be more defective and faulty because of their compactness. Scientists and engineers have created all of the essential components of nanocircuits such as transistors, logic gates and diodes. They have all been constructed from organic molecules, carbon nanotubes and nanowire semiconductors. The only thing left to do is find a way to eliminate the errors that come with such a small device and nanocircuits will become a way of all electronics. However, eventually there will be a limit as to how small nanocircuits can become and computers and electronics will reach their equilibrium speeds.

EXISTING PRODUCTS

Perhaps the most widely known product using nanocircuits would be the iPod nano. This electronic device is only the width of a pencil and no more than four inches long and two inches wide, yet it can hold several gigabytes of data letting the consumer practically put hundreds and even over a thousand songs on it. It utilizes flash memory which gives it the ability to store more data in a smaller size. Flash memory means it is a computer memory that has the ability to be erased, rewritten and reprogrammed. This can also be seen in the memory sticks used to transfer data from one place to another. One can take files from one computer, store them on the memory stick and transfer it to another computer. It is no more than a hard disk no bigger than a pack of gum. Because of the sheer size of these products, they must use nanocircuits in order to keep the size as small as it is.

Intel has become a leader in the computer industry when it comes to building smaller and faster microchips. One of their new products is the Intel Core2 Duo Processor. Although it is based on microarchitecture, it uses nanocircuitry in its design. They are up to 40% faster than microprocessors out now and are even more energy-efficient. Currently, the core duo contains over 150 million transistors, very characteristic of a nanocircuit which holds millions of transistors as well. The manufacturing process 65 nm, makes it a true nanocircuit. This is one of the leading processors for computers equipped for gaming and other high definition applications. Recently, Intel released information about a 45nm processor that will be released in 2007, which is slightly smaller than the 65 nm processor used now.

ECONOMIC IMPACT

With the vast improvements in reducing the size of circuits, comes a rising cost to produce these nano components. Scientists believe that one day a fabrication facility for making nanocircuit could cost as much as over $200 billion. The increased cost comes from the difficulty of producing such circuits as they take more time and effort than circuits today. The fabrication plant will create a raw nanocircuit-billions on billions of devices and wires whose functioning is rather limited. From the outside it will look like a lump of material with a handful of wires sticking out. Eventually the theory of Moore's Law will have to reach equilibrium with the fabrication methods currently used.

Circuits will only be able to be so fast and small without creating any severe problems. The cost for producing even better nanocircuits will increase further as more money will be needed to develop new fabrication methods and ways of designing faster, better nanocircuits. Until that time, companies like Intel will continue to thrive in the nano business with their promises of their chip being the fastest and better than their counterpart. Nanocircuits may still have their problems, but that will not stop companies from mass producing them in order to become the most technologically advanced company with the fastest product.

8

NANOTECHNOLOGY AND ITS IMPLICATIONS

Potential risks of nanotechnology can broadly be grouped into three areas:

- o the risk to health and environment from nanoparticles and nanomaterials;
- o the risk posed by molecular manufacturing (or advanced nanotechnology);
- o societal risks.

Nanoethics concerns the ethical and social issues associated with developments in nanotechnology, a science which encompass several fields of science and engineering, including biology, chemistry, computing, and materials science. Nanotechnology refers to the manipulation of very small-scale matter-a nanometer is one billionth of a meter, and nanotechnology is generally used to mean work on matter at 100 nanometers and smaller.

Social risks related to nanotechnology development include the possibility of military applications of nanotechnology (such as implants and other means for soldier enhancement) as well as enhanced surveillance capabilities through nano-sensors. However those applications still belong to science-fiction and will not be possible in the next decades. Significant environmental, health, and safety issues might arise with development in nanotechnology since some negative effects of nanoparticles in our environment might be overlooked. However nature itself creates all kinds of nanoobjects, so probable dangers are not due to the nanoscale alone, but due to the fact that toxic materials become more harmful when ingested or inhaled as nanoparticles. Note also that the early use of nanotechnology dates back over 2 milleniums. Nanotechnology has, for example, been used by the Romans since at least the fourth century AD to color glass by embedding gold nanoparticles, as evidenced by the Lycurgus cup. In discussing issues related to nanotechnology, the acronym NELSI is used to signify nanotechnology's ethical, legal, and social implications.

HEALTH AND SAFETY IMPLICATIONS

The mere presence of nanomaterials (materials that contain nanoparticles) is not in itself a threat. It is only certain aspects that can make them risky, in particular their mobility and their increased reactivity. Only if certain properties of certain nanoparticles were harmful to living beings or the environment would we be faced with a genuine hazard. In this case it can be called Nanopollution.

In addressing the health and environmental impact of nanomaterials we need to differentiate two types of nanostructures:

(1) Nanocomposites, nanostructured surfaces and nanocomponents (electronic, optical, sensors etc.), where nanoscale particles are incorporated into a substance, material or device ("fixed" nano-particles); and

(2) "free" nanoparticles, where at some stage in production or use individual nanoparticles of a

substance are present. These free nanoparticles could be nanoscale species of elements, or simple compounds, but also complex compounds where for instance a nanoparticle of a particular element is coated with another substance ("coated" nanoparticle or "core-shell" nanoparticle).

There seems to be consensus that, although one should be aware of materials containing fixed nanoparticles, the immediate concern is with free nanoparticles.

Because nanoparticles are very different from their everyday counterparts, their adverse effects cannot be derived from the known toxicity of the macro-sized material. This poses significant issues for addressing the health and environmental impact of free nanoparticles.

To complicate things further, in talking about nanoparticles it is important that a powder or liquid containing nanoparticles is almost never monodisperse, but will contain a range of particle sizes. This complicates the experimental analysis as larger nanoparticles might have different properties than smaller ones. Also, nanoparticles show a tendency to aggregate and such aggregates often behave differently from individual nanoparticles.

Health Issues

There are several potential entry routes for nanoparticles into the body. They can be inhaled, swallowed, absorbed through skin or be deliberately injected during medical procedures (or released from implants). Once within the body they are highly mobile and in some instances can even cross the blood-brain barrier. How these nanoparticles behave inside the organism is one of the big issues that needs to be resolved. The behavior of nanoparticles is a function of their size, shape and surface reactivity with the surrounding tissue.

They could cause overload on phagocytes, cells that ingest and destroy foreign matter, thereby triggering stress reactions that lead to inflammation and weaken the body's defense against other pathogens. Apart from what happens if non-degradable or slowly degradable nanoparticles accumulate in organs, another concern is their potential interaction with biological processes inside the body: because of their large surface, nanoparticles on exposure to tissue and fluids will immediately adsorb onto their surface some of the macromolecules they encounter. This may, for instance, affect the regulatory mechanisms of enzymes and other proteins.

Environmental Issues

Not enough data exists to know for sure if nanoparticles could have undesirable effects on the environment. Two areas are relevant here:

(1) In free form nanoparticles can be released in the air or water during production (or production accidents) or as waste byproduct of production, and ultimately accumulate in the soil, water or plant life.

(2) In fixed form, where they are part of a manufactured substance or product, they will ultimately have to be recycled or disposed of as waste. It is not known yet whether certain nanoparticles will constitute a completely new class of non-biodegradable pollutant. In case they do, it is not known how such pollutants could be removed from air or water because most traditional filters are not suitable for such tasks (their pores are too big to catch nanoparticles).

Health and environmental issues combine in the workplace of companies engaged in producing or using nanomaterials and in the laboratories engaged in nanoscience and nanotechnology research. It is safe to say that current workplace exposure standards for dusts cannot be applied directly to nanoparticle dusts.

To properly assess the health hazards of engineered nanoparticles the whole life cycle of these particles needs to be evaluated, including their fabrication, storage and distribution, application and potential abuse, and disposal. The impact on humans or the environment may vary at different stages of the life cycle.

Regarding the risks from molecular manufacturing, an often cited worst-case scenario is "grey goo", a hypothetical substance into which the surface of the earth might be transformed by self-replicating nanobots running amok. This concept has been analyzed by Freitas in "Some Limits to Global Ecophagy by Biovorous Nanoreplicators, with Public Policy Recommendations" With the advent of nan-biotech, a different scenario called green goo has been forwarded. Here, the malignant substance is not nanobots but rather self-replicating organisms engineered through nanotechnology.

A Need for Regulation

Regulatory bodies such as the Environmental Protection Agency and the Food and Drug Administration in the U.S. or the Health & Consumer Protection Directorate of the European Commission have started dealing with the potential risks posed by nanoparticles. So far, neither engineered nanoparticles nor the products and materials that contain them are subject to any special regulation regarding production, handling or labeling. The Material Safety Data Sheet that must be issued for certain materials often does not differentiate between bulk and nanoscale size of the material in question and even when it does these MSDS are advisory only. Studies of the health impact of airborne particles are the closest thing we have to a tool for assessing potential health risks from free nanoparticles. These studies have generally shown that the smaller the particles get, the more toxic they become. This is due in part to the fact that, given the same mass per volume, the dose in terms of particle numbers increases as particle size decreases.

Looking at all available data, it must be concluded that current risk assessment methodologies are not suited to the hazards associated with nanoparticles; in particular, existing toxicological and eco-toxicological methods are not up to the task; exposure evaluation (dose) needs to be expressed as quantity of nanoparticles and/or surface area rather than simply mass; equipment for routine detecting and measuring nanoparticles in air, water, or soil is inadequate; and very little is known about the physiological responses to nanoparticles.

Regulatory bodies in the U.S. as well as in the EU have concluded that nanoparticles form the potential for an entirely new risk and that it is necessary to carry out an extensive analysis of the risk. The challenge for regulators is whether a matrix can be developed which would identify nanoparticles and more complex nanoformulations which are likely to have special toxicological properties or whether it is more reasonable for each particle or formulation to be tested separately. Since null hypotheses are unfalsifiable in a true sense, nanoparticles and nanoformulations can never be proven safe.

A truly precautionary approach to regulation would severely impede development in the field of nanotechnology if we require safety studies for each and every nanoscience application. Consequently, the rush seems to be on to establish a research needs assessment in the nanocommunity to preclude universal safety studies.

While the outcome of these studies can form the basis for government and international regulations, a more reasonable approach might be development of a risk matrix that identifies likely culprits.

SOCIETAL IMPLICATIONS

Nanoethicists posit that such a transformative technology could exacerbate the divisions of rich and poor-the so-called "nano divide." However nanotechnology makes the production of technology, e.g. computers, cellular phones, health technology etcetera, cheaper and therefore accessible to the poor.

In fact, many of the most enthusiastic proponents of nanotechnology, such as transhumanists, see the nascent science as a mechanism to changing human nature itself-going beyond curing disease and enhancing human characteristics.

Discussions on nanoethics have been hosted by the federal government, especially in the context of "converging technologies"-a catch-phrase used to refer to nano, biotech, information technology, and cognitive science.

Possible Military Applications

Societal risks from the use of nanotechnology have also been raised. On the instrumental level, these include the possibility of military applications of nanotechnology (for instance, as in implants and other means for soldier enhancement like those being developed at the Institute for Soldier Nanotechnologies at MIT) as well as enhanced surveillance capabilities through nano-sensors.

There is also the possibility of nanotechnology being used to develop chemical weapons and because they will be able to develop the chemicals from the atom scale up, critics fear that chemical weapons developed from nano particles will be more dangerous than present chemical weapons.

Intellectual Property Issues

On the structural level, critics of nanotechnology point to a new world of ownership and corporate control opened up by nanotechnology. The claim is that, just as biotechnology's ability to manipulate genes went hand in hand with the patenting of life, so too nanotechnology's ability to manipulate molecules has led to the patenting of matter.

The last few years has seen a gold rush to claim patents at the nanoscale. Over 800 nano-related patents were granted in 2003, and the numbers are increasing year to year. Corporations are already taking out broad-ranging patents on nanoscale discoveries and inventions.

For example, two corporations, NEC and IBM, hold the basic patents on carbon nanotubes, one of the current cornerstones of nanotechnology. Carbon nanotubes have a wide range of uses, and look set to become crucial to several industries from electronics and computers, to strengthened materials to drug delivery and diagnostics.

Carbon nanotubes are poised to become a major traded commodity with the potential to replace major conventional raw materials. However, as their use expands, anyone seeking to manufacture or sell carbon nanotubes, no matter what the application, must first buy a license from NEC or IBM.

BENEFITS AND RISKS

Nanotechnologies may provide new solutions for the millions of people in developing countries who lack access to basic services, such as safe water, reliable energy, health care, and education. The United Nations has set Millennium Development Goals for meeting these needs. The 2004 UN Task Force on Science, Technology and Innovation noted that some of the advantages of nanotechnology include production using little labor, land, or maintenance, high productivity, low cost, and modest requirements for materials and energy. Many developing countries, for example Costa Rica, Chile, Bangladesh, Thailand, and Malaysia, are investing considerable resources in research and development of nanotechnologies.

Emerging economies such as Brazil, China, India and South Africa are spending millions of US dollars annually on R&D, and are rapidly increasing their scientific output as demonstrated by their increasing numbers of publications in peer-reviewed scientific publications.

Potential opportunities of nanotechnologies to help address critical international development priorities include improved water purification systems, energy systems, medicine and pharmaceuticals, food production and nutrition, and information and communications technologies.

Nanotechnologies are already incorporated in products that are on the market. Other nanotechnologies are still in the research phase, while others are concepts that are years or decades away from development.

Applying nanotechnologies in developing countries raises similar questions about the environmental, health, and societal risks described in the previous section. Additional challenges have been raised regarding the linkages between nanotechnology and development.

Protection of the environment, human health and worker safety in developing countries often suffers from a combination of factors that can include but are not limited to lack of robust environmental, human health, and worker safety regulations; poorly or unenforced regulation which is linked to a lack of physical (e.g., equipment) and human capacity (i.e., properly trained regulatory staff).

Often, these nations require assistance, particularly financial assistance, to develop the scientific and institutional capacity to adequately assess and manage risks, including the necessary infrastructure such as laboratories and technology for detection.

Very little is known about the risks and broader impacts of nanotechnology. At a time of great uncertainty over the impacts of nanotechnology it will be challenging for governments, companies, civil society organizations, and the general public in developing countries, as in developed countries, to make decisions about the governance of nanotechnology.

Companies, and to a lesser extent governments and universities, are receiving patents on nanotechnology. The rapid increase in patenting of nanotechnology is illustrated by the fact that in the US, there were 500 nanotechnology patent applications in 1998 and 1,300 in 2000.

Some patents are very broadly defined, which has raised concern among some groups that the rush to patent could slow innovation and drive up costs of products, thus reducing the potential for innovations that could benefit low income populations in developing countries.

There is a clear link between commodities and poverty. Many least developed countries are dependent on a few commodities for employment, government revenue, and export earnings. Many applications of nanotechnology are being developed that could impact global demand for specific commodities.

For instance, certain nanoscale materials could enhance the strength and durability of rubber, which might eventually lead to a decrease in demand for natural rubber. Other nanotechnology applications may result in increases in demand for certain commodities.

For example, demand for titanium may increase as a result of new uses for nanoscale titanium oxides, such as titanium dioxide nanotubes that can be used to produce and store hydrogen for use as fuel.

Various organizations have called for international dialogue on mechanisms that will allow developing countries to anticipate and proactively adjust to these changes.

In 2003, Meridian Institute began the Global Dialogue on Nanotechnology and the Poor: Opportunities and Risks (GDNP) to raise awareness of the opportunities and risks of nanotechnology for developing countries, close the gaps within and between sectors of society to catalyze actions that address specific opportunities and risks of nanotechnology for developing countries, and identify ways that science and technology can play an appropriate role in the development process.

The GDNP has released several publicly accessible papers on nanotechnology and development, including "Nanotechnology and the Poor: Opportunities and Risks-Closing the Gaps Within and Between Sectors of Society"; "Nanotechnology, Water, and Development"; and "Overview and Comparison of Conventional and Nano-Based Water Treatment Technologies".

Studies on the Implications of Nanotechnology:

- o In July 2003 the US Environmental Protection Agency issued the first research solicitation in the area of nanotechnology implications, "Exploratory Research to Anticipate Future Environmental

Issues-Part 2: Impacts of Manufactured Nanomaterials on Human Health and the Environment." In September 2004 US EPA partnered with the National Science Foundation and the Centers for Disease Control to issue a second research solicitation, "Nanotechnology Research Grants Investigating Environmental and Human Health Effects of Manufactured Nanomaterials: A Joint Research Solicitation-EPA, NSF, NIOSH."

- In August 2005, a task force consisting of 50+ international experts from various fields was organized by the Center for Responsible Nanotechnology to study the societal implications of molecular nanotechnology.
- In October 2005, the National Science Foundation announced that it would fund two national centers to research the potential societal implications of nanotechnology. Located at the University of California, Santa Barbara and Arizona State University, researchers at these two centers are exploring a wide range of issues including nanotechnology's historical context, technology assessment, innovation and globalization issues, and societal perceptions of risk.
- A book by Geoffrey Hunt and Michael Mehta (2006) entitled Nanotechnology: Risk, Ethics and Law (London: Earthscan Book) provides a global overview of the state of nanotechology and society in Europe, the USA, Japan and Canada, and examines the ethics, the environmental and public health risks, and the governance and regulation of this most promising, and potentially most dangerous, of all technologies.

- o Determining a set of pathways for the development of molecular nanotechnology is now an objective of a broadly based technology roadmap project led by Battelle (the manager of several U.S. National Laboratories) and the Foresight Institute. That roadmap should be completed by early 2007.
- o In October 2006, the International Council on Nanotechnology (ICON) based at Rice University published a survey of nanomaterial handling practices being used by industrial and academic workplaces on four continents. The survey revealed that more information is needed to protect against the potential occupational risks associated with handling free nanoparticles. ICON also maintains the Virtual Journal of Nanotechnology Environment, Health & Safety (VJ-NanoEHS) which is a compilation of citations to peer-reviewed studies on risk issues.

9

THE MOLECULAR NANOTECHNOLOGY

Molecular nanotechnology (MNT) is the engineering of functional systems at the molecular scale. An equivalent definition would be "machines at the molecular scale designed and built atom-by-atom". This is distinct from nanoscale materials. Based on Richard Feynman's vision of miniature factories using nanomachines to build complex products (including additional nanomachines), this advanced form of nanotechnology (or molecular manufacturing) would make use of positionally-controlled mechanosynthesis guided by molecular machine systems. MNT would involve combining physical principles demonstrated by chemistry, other nanotechnologies, and the molecular machinery of life with the systems engineering principles found in modern macroscale factories. Its most well-known exposition is in the books of K. Eric Drexler particularly Engines of Creation.

Formulating a roadmap for the development of MNT is now an objective of a broadly based technology roadmap project led by Battelle (the manager of several U.S. National Laboratories) and the Foresight Institute. The roadmap should be completed by early 2007. In August 2005, a task force consisting of 50+ international experts from various fields was organized by the Center for

Responsible Nanotechnology to study the societal implications of molecular nanotechnology.

While conventional chemistry employs stochastic processes driven toward some equilibrium to obtain stochastic results, and biology exploits stochastic processes to obtain deterministic results based on complex enzyme-catalyzed reaction chains optimized through billions of years of evolutionary feedback, molecular nanotechnology would employ novel (and as yet unspecified) deterministic nanoscale processes to obtain deterministic results. The desire in molecular nanotechnology would be to place molecular moieties in deterministic locations with deterministic orientation to obtain desired chemical reactions, and then to build systems by further assembling the products of these reactions.

Ralph Merkle has compared today's manufacturing methods (in contrast to mechanosynthesis) to an attempt to build interesting Lego brick constructions while wearing boxing gloves: "Casting, grinding, milling and even lithography move atoms in great thundering statistical herds. It's like trying to make things out of LEGO blocks with boxing gloves on your hands. Yes, you can push the LEGO blocks into great heaps and pile them up, but you can't really snap them together the way you'd like." It has been posited that molecular nanotechnology could offer much cleaner manufacturing processes than today's bulk technology.

BACKGROUND

Smart Materials and Nanosensors: One proposed application of MNT is the development of so-called smart materials. This term refers to any sort of material designed and engineered at the nanometer scale to perform a specific task, and encompasses a wide variety of possible commercial applications. One example is materials designed to respond differently to various molecules; such a capability could lead, for example, to artificial drugs which would recognize and render inert specific viruses. Another is the idea of self-healing structures, which would repair small tears in a surface naturally in the same way as self-sealing tires or human skin; and while this technology is relatively new, it is already seeing commercial application in various engineering plastics.

A nanosensor created by MNT would resemble a smart material, involving a small component within a larger machine that would react to its environment and change in some fundamental, intentional way. As a very simple example: a photosensor could passively measure the incident light and discharge its absorbed energy as electricity when the light passes above or below a specified threshold, sending a signal to a larger machine. Such a sensor would cost less and use less power than a conventional sensor, and yet function usefully in all the same applications-for example, turning on parking lot lights when it gets dark.

While smart materials and nanosensors both exemplify useful applications of MNT, they pale in comparison with the complexity of the technology most popularly associated with the term: the replicating nanorobot.

Replicating Nanorobots

MNT nanofacturing is popularly linked with the idea of swarms of coordinated nanoscale robots working together, as proposed by Drexler in his 1986 popular discussions of the subject. It is proposed that sufficiently capable nanorobots could construct more nanorobots.

However, critics doubt both the feasibility of self-replicating nanorobots and the feasibilty of control if self-replicating nanorobots could be achieved: they cite the possibility of mutations removing any control and favoring reproduction of mutant pathogenic variations. Advocates address the second doubt by arguing that bacteria are (of necessity) evolved to evolve, while nanorobot mutation can be actively prevented by common error-correcting techniques. Similar ideas are advocated in the Foresight Guidelines on Molecular Nanotechnology.

Recent technical proposals for MNT nanofactories do not include self-replicating nanorobots, and recent ethical guidelines prohibit self-replication.

Medical Nanorobots

One of the most important applications of MNT would be medical nanorobotics or nanomedicine, an area pioneered by Robert Freitas in numerous books and papers. The ability to design, build, and deploy large numbers of medical nanorobots would, at an optimum, make possible the rapid elimination of disease and the reliable and relatively painless recovery from physical trauma. Medical nanorobots might also make possible the convenient correction of genetic defects, and help to ensure a greatly expanded health-span. More controversially, medical nanorobots might be used to augment natural human capabilities. However, mechanical medical nanodevices would not be allowed (or designed) to self-replicate inside the human body, nor would medical nanorobots have any need for self-replication themselves since they would be manufactured exclusively in carefully regulated nanofactories.

Utility Fog

Another proposed application of nanotechnology involves utility fog-in which a cloud of networked microscopic robots (simpler than assemblers) changes its shape and properties to form macroscopic objects and tools in accordance with software commands. Rather than modify the current practices of consuming material goods in different forms, utility fog would simply replace many physical objects.

Phased-Array Optics

Yet another proposed application would be phased-array optics (PAO). PAO would used the principle of phased-array millimeter technology but at optical wavelengths. This would permit the duplication of any sort of optical effect but virtually. Users could request holograms, sunrises and sunsets, or floating lasers as the mood strikes. PAO systems were described in BC Crandall's Nanotechnology: Molecular Speculations on Global Abundance in the Brian Wowk article "Phased-Array Optics".

THE SOCIAL IMPACTS

Despite the current early developmental status of nanotechnology and molecular nanotechnology, much concern surrounds MNT's anticipated impact on economics and on law. Some conjecture that MNT would elicit a strong public-opinion backlash, as has occurred recently around genetically modified plants and the prospect of human cloning. Whatever the exact effects, MNT, if achieved, would tend to upset existing economic structures by reducing the scarcity of manufactured goods and making many more goods (such as food and health aids) manufacturable.

It is generally considered that future citizens of a molecular-nanotechnological society would still need money, in the form of unforgeable digital cash or physical specie (in special circumstances). They might use such money to buy goods and services that are unique, or limited within the solar system. These might include: matter, energy, information, real estate, design services, entertainment services, legal services, fame, political power, or the attention of other people to your political/ religious/philosophical message. Furthermore, futurists must consider war, even between prosperous states, and non-economic goals.

If MNT were realized, some resources would remain limited, because unique physical objects are limited (a plot of land in the real Jerusalem, mining rights to the larger near-earth asteroids) or because they depend on the goodwill of a particular person (the love of a famous person, a painting from a famous artist). Demand will always exceed supply for some things, and a political economy may continue to exist in any case. Whether the interest in these limited resources would diminish with the advent of virtual reality, where they could be easily substituted, is yet unclear; one reason why it might not is a hypothetical preference for "the real thing".

RISKS

Molecular nanotechnology is one of the technologies that some analysts believe could lead to a Technological Singularity. Some feel that molecular nanotechnology would have daunting risks. It conceivably could enable cheaper and more destructive conventional weapons. Also, molecular nanotechnology might permit weapons of mass destruction that could self-replicate, as viruses and cancer cells do when attacking the human body. Commentators generally agree that, in the event molecular nanotechnology were developed, mankind should permit self-replication only under very controlled or "inherently safe" conditions.

A fear exists that nanomechanical robots, if achieved, and if designed to self-replicate using naturally occurring materials (a difficult task), could consume the entire planet in their hunger for raw materials, or simply crowd out natural life, out-competing it for energy (as happened historically when blue-green algae appeared and outcompeted earlier life forms). Some commentators have referred to this situation as the "grey goo" or "ecophagy" scenario. K. Eric Drexler considers an accidental "grey goo" scenario extremely unlikely and says so in later editions of Engines of Creation. The "grey goo" scenario begs the Tree Sap Answer: what chances exist that one's car could spontaneously mutate into a wild car, run off-road and live in the forest off tree sap?

In light of this perception of potential danger, the Foresight Institute (founded by K. Eric Drexler to prepare for the arrival of future technologies) has drafted a set of guidelines for the ethical development of nanotechnology. These include the banning of free-foraging self-replicating pseudo-organisms on the Earth's surface, at least, and possibly in other places.

TECHNICAL ISSUES AND THE CRITICISMS

A section heading in Drexler's Engines of Creation reads "Universal Assemblers", and the following text speaks of molecular assemblers which could hypothetically "build almost anything that the laws of nature allow to exist." Drexler's colleague Ralph Merkle has noted that, contrary to widespread legend,, Drexler never claimed that assembler systems could build absolutely any molecular structure. The endnotes in Drexler's book explain the qualification "almost": "For example, a delicate structure might be designed that, like a stone arch, would self-destruct unless all its pieces were already in place. If there were no room in the design for the placement and removal of a scaffolding, then the structure might be impossible to build. Few structures of practical interest seem likely to exhibit such a problem, however." In 1992, Drexler published Nanosystems: molecular machinery, manufacturing, and computation, a detailed proposal for synthesizing stiff, diamond-based structures using a table-top factory. Although such a nanofactory would be far less powerful than a protean universal assembler, it would still be enormously capable. Diamondoid structures and other stiff covalent structures, if achieved, would have a wide range of possible applications, going far beyond current MEMS technology. However, no proposal was put forward for building the table-top factory in the absence of a near-universal assembler.

Several researchers, including Dr. Smalley, have attacked the notion of universal assemblers, leading to a rebuttal from Drexler and colleagues, and eventually to an exchange of letters. Smalley argues that chemistry is extremely complicated, reactions are hard to control, and that a universal assembler is science fiction. Drexler and colleagues, however, note that Drexler never proposed universal assemblers able to make absolutely anything, but had instead proposed more limited assemblers able to make a very wide variety of things. They challenge the relevance of Smalley's arguments to the more specific proposals advanced in Nanosystems.

FEASIBILITY OF THE PROPOSALS

The feasibility of Drexler's proposals largely depends, therefore, on whether designs like those in Nanosystems could be built in the absence of a universal assembler to build them and would work as described. Supporters of molecular nanotechnology frequently claim that no significant errors have been discovered in Nanosystems since 1992. Even some critics concede that "Drexler has carefully considered a number of physical principles underlying the 'high level' aspects of the nanosystems he proposes and, indeed, has thought in some detail" about some issues.

Other critics claim, however, that Nanosystems omits important chemical details about the low-level 'machine language' of molecular nanotechnology (Smalley, Atkinson, Moriarty, Jones). They also claim that much of the other low-level chemistry in Nanosystems requires extensive further work, and that Drexler's higher-level designs therefore rest on speculative foundations.

Drexler argues that we may need to wait until our conventional nanotechnology improves before solving these issues: "Molecular manufacturing will result from a series of advances in molecular machine systems, much as the first Moon landing resulted from a series of advances in liquid-fuel rocket systems. We are now in a position like that of the British Interplanetary Society of the 1930s which described how multistage liquid-fueled rockets could reach the Moon and pointed to early rockets as illustrations of the basic principle."

To summarize the arguments against feasibility, a primary barrier to achieving molecular nanotechnology is the lack of an efficient way to create machines on a molecular/atomic scale, especially in the absence of a well-defined path toward a self-replicating assembler.

A second difficulty in reaching molecular nanotechnology is design. Hand design of a gear or bearing at the level of atoms is a grueling task. While Drexler, Merkle and others have created a few designs of simple parts, no comprehensive design effort for anything approaching the complexity of a Model T Ford has been attempted.

A third difficulty in achieving molecular technology is separating successful trials from failures, and elucidating the failure mechanisms of the failures. Unlike biological evolution, which proceeds by random variations in ensembles of organisms combined with deterministic reproduction/extinction as a selection process to achieve great complexity after billions of years (a set of mechanisms which Richard Dawkins has referred to as a "blind watchmaker"), deliberate design and building of nanoscale mechanisms requires a means other than reproduction/extinction to winnow successes from failures in proceeding from simplicity to complexity. Such means are difficult to provide (and presently non-existent) for anything other than small assemblages of atoms viewable by an AFM or STM.

Thus, even in the latest report A Matter of Size: Triennial Review of the National Nanotechnology Initiative put out by the National Academies Press in December 2006, (roughly twenty years after Engines of Creation was published) no clear way forward toward molecular nanotechnology is seen, as per the conclusion on page 108 of that report: "Although theoretical calculations can be made today, the eventually attainable range of chemical reaction cycles, error rates, speed of operation, and thermodynamic efficiencies of such bottom-up manufacturing systems cannot be reliably predicted at this time.

Thus, the eventually attainable perfection and complexity of manufactured products, while they can be calculated in theory, cannot be predicted with confidence. Finally, the optimum research paths that might lead to systems which greatly exceed the thermodynamic efficiencies and other capabilities of biological systems cannot be reliably predicted at this time. Research funding that is based on the ability of investigators to produce experimental demonstrations that link to abstract models and guide long-term vision is most appropriate to achieve this goal." Perhaps the eventual "Technology Roadmap for Productive Nanosystems" will exhibit a more hopeful tone.

It is perhaps interesting to ask whether or not most structures consistent with physical law can in fact be manufactured. But such a question is a great deal more difficult to answer than, for example, the four-color map theorem which was proposed in 1852 and proven in 1976, and it is conceptually impossible to prove the negative of this question since no proof by counter-example can be provided. In any case, as Richard Feynman once said, "It is scientific only to say what's more likely or less likely, and not to be proving all the time what's possible or impossible."

The tool-tips modelled in this work are intended to be used only in carefully controlled environments (e.g., vacuum). Maximum acceptable limits for tooltip translational and rotational misplacement errors are reported in paper III--tool-tips must be positioned with great accuracy to avoid bonding the dimer incorrectly. However, a skeptical observer may look at the positional uncertainty of carbon atom placement of that work and conclude that it is achieved only via a simple cheat as per the text of the article: "Simulations were performed by tethering all 50 carbon atoms in the topmost plane of the tool handle to their energy-minimized positions using a large force restraint equal to the MM2 force field C-C bond stiffness of 440 N/m, or 633 kcal/mol-Å, with different initial atomic positions and randomized initial velocities for each independent simulation...." While this approach certainly minimizes computation complexity, it unrealistically ignores the need for some type of frame to position the tool handle with respect to the workpiece, the frame of necessity having its own non-infinite stiffness and finite vibrational modes leading to additional positional uncertainty. Thus the authors are over-optimistic in their predictions of positional uncertainty, and analysis of a design having a supporting frame may reveal that there is no way to achieve the necessary positional uncertainty at 300K, 80K, or even 20K.

Over 100,000 CPU hours were invested in this latest study. The DCB6 tooltip motif, initially described at a Foresight Conference in 2002, was the first complete tooltip ever proposed for diamond mechanosynthesis and remains the only tooltip motif that has been successfully simulated for its intended function on a full 200-atom diamond surface.

Further research to consider additional tool-tips will require time-consuming computational chemistry and difficult laboratory work.

A working nanofactory would require a variety of well-designed tips for different reactions, and detailed analyses of placing atoms on more complicated surfaces. Although this appears a challenging problem given current resources, many tools will be available to help future researchers: Moore's Law predicts further increases in computer power, semiconductor fabrication techniques continue to approach the nanoscale, and researchers grow ever more skilled at using proteins, ribosomes and DNA to perform novel chemistry.

10

NANOPOLLUTION AND ITS SOCIAL EFFECTS

Nanopollution is a generic name for all waste generated by nanodevices or during the nanomaterials manufacturing process. This kind of waste may be very dangerous because of its size. It can float in air the and might easily penetrate animal and plant cells causing unknown effects. Most human made nanoparticles do not appear in nature so living organism may not have appropriate means to deal with this kind of waste. It is probably one great challenge to nanotechnology: How to deal with its nanopollutants and nanowaste.

THE LAW OF MOORE

Moore's Law is the empirical observation made in 1965 that the number of transistors on an integrated circuit for minimum component cost doubles every 24 months. It is attributed to Gordon E. Moore (born 1929), a co-founder of Intel. Although it is sometimes quoted as every 18 months, Intel's official Moore's Law page, as well as an interview with Gordon Moore himself, state that it is every two years.

PRIMARY FORMS

The term Moore's Law was coined by Carver Mead around 1970. Moore's original statement can be found in his publication "Cramming more components onto integrated circuits", Electronics Magazine 19 April 1965:

"The complexity for minimum component costs has increased at a rate of roughly a factor of two per year... Certainly over the short term this rate can be expected to continue, if not to increase. Over the longer term, the rate of increase is a bit more uncertain, although there is no reason to believe it will not remain nearly constant for at least 10 years. That means by 1975, the number of components per integrated circuit for minimum cost will be 65,000. I believe that such a large circuit can be built on a single wafer."

Under the assumption that chip "complexity" is proportional to the number of transistors, regardless of what they do, the law has largely held the test of time to date. However, one could argue that the per-transistor complexity is less in large RAM cache arrays than in execution units. From this perspective, the validity of one formulation of Moore's Law may be more questionable.

Gordon Moore's observation was not named a "law" by Moore himself, but by the Caltech professor, VLSI pioneer, and entrepreneur Carver Mead. Moore, indicating that it cannot be sustained indefinitely, has since observed "It can't continue forever. The nature of exponentials is that you push them out and eventually disaster happens."

Moore may have heard Douglas Engelbart, a co-inventor of today's mechanical computer mouse, discuss the projected downscaling of integrated circuit size in a 1960 lecture. In 1975, Moore projected a doubling only every two years. He is adamant that he himself never said "every 18 months", but that is how it has been quoted. The SEMATECH roadmap follows a 24 month cycle.

Understanding the Law

Moore's law is not about just the density of transistors that can be achieved, but about the density of transistors at which the cost per transistor is the lowest. As more transistors are made on a chip the cost to make each transistor reduces but the chance that the chip will not work due to a defect rises. If the rising cost of discarded non working chips is balanced against the reducing cost per transistor of larger chips, then as Moore observed in 1965 there is a number of transistors or complexity at which "a minimum cost" is achieved. He further observed that as transistors were made smaller through advances in photolithography this number would increase "a rate of roughly a factor of two per year".

Formulations of the Law

The most popular formulation is of the doubling of the number of transistors on integrated circuits every 18 months. At the end of the 1970s, Moore's Law became known as the limit for the number of transistors on the most complex chips. However, it is also common to cite Moore's Law to refer to the rapidly continuing advance in computing power per unit cost, because increase in transistor count is also a rough measure of computer processing power.

A similar law (sometimes called Kryder's Law) has held for hard disk storage cost per unit of information. The rate of progression in disk storage over the past decades has actually sped up more than once, corresponding to the utilization of error correcting codes, the magnetoresistive effect and the giant magnetoresistive effect. The current rate of increase in hard drive capacity is roughly similar to the rate of increase in transistor count. However, recent trends show that this rate is dropping, and has not been met for the last three years.

See Hard disk capacity. Another version states that RAM storage capacity increases at the same rate as processing power.

THE DRIVER OF INDUSTRY

Although Moore's Law was initially made in the form of an observation and forecast, the more widely it became accepted, the more it served as a goal for an entire industry. This drove both marketing and engineering departments of semiconductor manufacturers to focus enormous energy aiming for the specified increase in processing power that it was presumed one or more of their competitors would soon actually attain. In this regard, it can be viewed as a self-fulfilling prophecy.

The implications of Moore's Law for computer component suppliers are very significant. A typical major design project (such as an all-new CPU or hard drive) takes between two and five years to reach production-ready status. In consequence, component manufacturers face enormous timescale pressures-just a few weeks of delay in a major project can spell the difference between great success and massive losses, even bankruptcy. Expressed as "a doubling every 18 months", Moore's Law suggests the phenomenal progress of technology in recent years.

Expressed on a shorter timescale, however, Moore's Law equates to an average performance improvement in the industry as a whole of close to 1% per week. For a manufacturer competing in the competitive CPU market, a new product that is expected to take three years to develop and is just three or four months late is 10 to 15% slower, bulkier, or lower in storage capacity than the directly competing products, and is usually unsellable. (If instead we accept that performance doubles every 24 months, rather than every 18 months, a 3 to 4 month delay would mean 8 to 11% less performance.)

Future Trends

As of Q1 2007, most PC processors are currently fabricated on a 65nm process, with some 90 nm chips still left in retail channels, mostly from AMD, as they are slightly behind Intel in transitioning away from 90 nm. On January 27, 2007, Intel demonstrated a working 45nm chip which they intend to begin mass-producing in late 2007. This new family of chips has been given the codename "Penryn". A decade ago, chips were built using a 500 nm process. Companies are working on using nanotechnology to solve the complex engineering problems involved in producing chips at the 30 nm and smaller levels-a process that will postpone the industry meeting the limits of Moore's Law.

Recent computer industry technology "roadmaps" predict (as of 2001) that Moore's Law will continue for several chip generations. Depending on the doubling time used in the calculations, this could mean up to 100 fold increase in transistor counts on a chip in a decade. The semiconductor industry technology roadmap uses a three-year doubling time for microprocessors, leading to about ninefold increase in a decade.

In early 2006, IBM researchers announced that they had developed a technique to print circuitry only 29.9 nm wide using deep-ultraviolet (DUV, 193-nanometer) optical lithography. IBM claims that this technique may allow chipmakers to use current methods for seven years while continuing to achieve results predicted by Moore's Law. New methods that can achieve smaller circuits are predicted to be substantially more expensive.

Since the rapid exponential improvement could (in theory) put 100 GHz personal computers in every home and 20 GHz devices in every pocket, some commentators have speculated that sooner or later computers will meet or exceed any conceivable need for computation. This is only true for some problems-there are others where exponential increases in processing power are matched or exceeded by exponential increases in complexity as the problem size increases. See computational complexity theory and complexity classes P and NP for a (somewhat theoretical) discussion of such problems, which occur very commonly in applications such as scheduling.

The exponential increase in frequency of operation as the only method of increasing computation speed is misleading. What matters is the exponential increase in useful work (or instructions) executed per unit time. In fact, newer processors are actually being made at lower clock speeds, with focus on larger caches and multiple computing cores. This occurs because higher clock speeds correspond to exponential increases in temperature, making it possible to have a CPU that is capable of running at 4.1 GHz for only a couple hundred dollars (using practical, yet uncommon methods of cooling), but it is almost impossible to produce a CPU that runs reliably at speeds higher than 4.3 GHz or so.

Extrapolation partly based on Moore's Law has led futurists such as Vernor Vinge, Bruce Sterling, and Ray Kurzweil to speculate about a technological singularity. However, on April 13, 2005, Gordon Moore himself stated in an interview that the law may not hold for too long, since transistors may reach the limits of miniaturization at atomic levels. "In terms of size [of transistor] you can see that we're approaching the size of atoms which is a fundamental barrier, but it'll be two or three generations before we get that far-but that's as far out as we've ever been able to see. We have another 10 to 20 years before we reach a fundamental limit. By then they'll be able to make bigger chips and have transistor budgets in the billions." While this time horizon for Moore's Law scaling is possible, it does not come without underlying engineering challenges. One of the major challenges in integrated circuits that use nanoscale transistors is increase in parameter variation and leakage currents.

As a result of variation and leakage, the design margins available to do predictive design is becoming harder and additionally such systems dissipate considerable power even when not switching. Adaptive and statistical design along with leakage power reduction is critical to sustain scaling of CMOS. A good treatment of these topics is covered in Leakage in Nanometer CMOS Technologies. Other scaling challenges include:

1. The ability to control parasitic resistance and capacitance in transistors,
2. The ability to reduce resistance and capacitance in electrical interconnects,
3. The ability to maintain proper transistor electrostatics that allow the gate terminal to control the ON/OFF behavior,
4. Increasing effect of line edge roughness,
5. Dopant fluctuations,
6. System level power delivery,
7. Thermal design to effectively handle the dissipation of delivered power, and
8. Solve all these challenges with ever-reducing cost of manufacturing of the overall system.

Kurzweil projects that a continuation of Moore's Law until 2019 will result in transistor features just a few atoms in width. Although this means that the strategy of ever finer photolithography will have run its course, he speculates that this does not mean the end of Moore's Law:

"Moore's Law of Integrated Circuits was not the first, but the fifth paradigm to provide accelerating price-performance. Computing devices have been consistently multiplying in power (per unit of time) from the mechanical calculating devices used in the 1890 US Census, to Turing's relay-based "Robinson" machine that cracked the German Enigma code, to the CBS vacuum tube computer that predicted the election of Eisenhower, to the transistor-based machines used in the first space launches, to the integrated-circuit-based personal [computers].

Thus, Kurzweil conjectures that it is likely that some new type of technology will replace current integrated-circuit technology, and that Moore's Law will hold true long after 2020. He believes that the exponential growth of Moore's Law will continue beyond the use of integrated circuits into technologies that will lead to the technological singularity. The Law of Accelerating Returns described by Ray Kurzweil has in many ways altered the public's perception of Moore's Law. It is a common (but mistaken) belief that Moore's Law makes predictions regarding all forms of technology, when it actually only concerns semiconductor circuits. Many futurists still use the term "Moore's Law" to describe ideas like those put forth by Kurzweil.

Krauss and Starkman announced an ultimate limit of around 600 years in their paper "Universal Limits of Computation", based on rigorous estimation of total information-processing capacity of any system in the Universe. Then again, the law has often met obstacles that appeared insurmountable, before soon surmounting them. In that sense, Mr. Moore says he now sees his law as more beautiful than he had realised. "Moore's Law is a violation of Murphy's Law. Everything gets better and better."

Not all aspects of computing technology develop in capacities and speed according to Moore's Law. Random Access Memory (RAM) speeds and hard drive seek times improve at best a few percentage points each year. Since the capacity of RAM and hard drives is increasing much faster than is their access speed, intelligent use of their capacity becomes more and more important. It now makes sense in many cases to trade space for time, such as by precomputing indexes and storing them in ways that facilitate rapid access, at the cost of using more disk and memory space: space is getting cheaper relative to time.

Another, sometimes misunderstood, point is that exponentially improved hardware does not necessarily imply exponentially improved software to go with it. The productivity of software developers most assuredly does not increase exponentially with the improvement in hardware, but by most measures has increased only slowly and fitfully over the decades. Software tends to get larger and more complicated over time, and Wirth's law even states that "Software gets slower faster than hardware gets faster".

Moreover, there is popular misconception that the clock speed of a processor determines its speed, also known as the Megahertz Myth. This actually also depends on the number of instructions per tick which can be executed (as well as the complexity of each instruction), and so the clock speed can only be used for comparison between two identical circuits. Of course, other factors must be taken into consideration such as the bus size and speed of the peripherals. Therefore, most popular evaluations of "computer speed" are inherently biased, without an understanding of the underlying technology. This was especially true during the Pentium era when popular manufacturers played with public perceptions of speed, focusing on advertising the clock rate of new products. It is also important to note that transistor density in multi-core CPUs does not necessarily reflect a similar increase in practical computing power, due to the unparallelized nature of most applications.

As the cost to the consumer of computer power falls, the cost for producers to achieve Moore's Law has the opposite trend: R&D, manufacturing, and test costs have increased steadily with each new generation of chips. As the cost of semiconductor equipment is expected to continue increasing, manufacturers must sell larger and larger quantities of chips to remain profitable. (The cost to tape-out a chip at 180 nm was roughly $300,000 USD. The cost to tape-out a chip at 90 nm exceeds $750,000 USD, and the cost is expected to exceed $1.0M USD for 65 nm.) In recent years, analysts have observed a decline in the number of "design starts" at advanced process nodes (130 nm and below.)

While these observations were made in the period after the 2000 economic downturn, the decline may be evidence that traditional manufacturers in the long-term global market cannot economically sustain Moore's Law. However, Intel was reported in 2005 as stating that the downsizing of silicon chips with good economics can continue for the next decade. Intel's prediction of increasing use of materials other than silicon, was verified in mid-2006, as was its intent of using trigate transistors around 2009. Researchers from IBM and Georgia Tech created a new speed record when they ran a silicon/germanium helium supercooled transistor at 500 gigahertz (GHz). The transistor operated above 500 GHz at 4.5 K (-451°F) and simulations showed that it could likely run at 1 THz (1,000 GHz), although this was only a single transistor, and practical desktop CPUs running at this speed are extremely unlikely using contemporary silicon chip techniques.

11

EXPONENTIAL GROWTH

In mathematics, exponential growth (or geometric growth) occurs when the growth rate of a function is always proportional to the function's current size. Such growth is said to follow an exponential law (but see also Malthusian growth model). This implies for any exponentially growing quantity, the larger the quantity gets, the faster it grows. But it also implies that the relationship between the size of the dependent variable and its rate of growth is governed by a strict law, of the simplest kind: direct proportion. It is proved in calculus that this law requires that the quantity is given by the exponential function, if we use the correct time scale. This explains the name.

THE INTUITION

The phrase exponential growth is often used in nontechnical contexts to mean merely surprisingly fast growth. In a strictly mathematical sense, though, exponential growth has a precise meaning and does not necessarily mean that growth will happen quickly. In fact, a population can grow exponentially but at a very slow absolute rate (as when money in a bank account earns a very low interest rate, for instance), and can grow surprisingly fast without growing exponentially. And some functions, such as the logistic function, approximate exponential growth over only part of their range. The "technical details" section below explains exactly what is required for a function to exhibit true exponential growth.

Limitations of Exponential Models

As discussed above, an important point about exponential growth is that even when it seems slow on the short run, it becomes impressively fast on the long run, with the initial quantity doubling at the doubling time, then doubling again and again. For instance, a population growth rate of 2% per year may seem small, but it actually implies doubling after 35 years, doubling again after another 35 years (i.e. becoming 4 times the initial population), etc. This implies that both the observed quantity and its time derivative will become several orders of magnitude larger than what was initially meant by the person who conceived the growth model. Because of this, some effects not initially taken into account will distort the growth law, usually moderating it as for instance in the logistic law. Exponential growth of a quantity placed in the real world (i.e. not in the abstract world of mathematics) is a model valid for a temporary period of time only.

For this reason, some people challenge the exponential growth model on the ground that it is valid for the short term only, i.e. nothing can grow indefinitely. For instance, a population in a closed environment cannot continue growing if it eats up all the available food and resources; industry cannot continue pumping carbon from the underground into the atmosphere beyond the limits connected with oil reservoirs and the consequences of climate change; etc. Problems of this kind exist for every mathematical representation of the real world, but are specially felt for exponential growth, since with this model growth accelerates as variables increase in a positive feedback, to a point were human response time to inconvenients can be insufficient (on these points, see also the Exponential stories below).

Exponential Growth

- o Biology.
- o Microorganisms in a culture dish will grow exponentially, at first, after the first microorganism appears (but then logistically until the available food is exhausted, when growth stops).
- o A virus (SARS, West Nile, smallpox) of sufficient infectivity ($k > 0$) will spread exponentially at first, if no artificial immunization is available. Each infected person can infect multiple new people.
- o Human population, if the number of births and deaths per person per year were to remain at current levels (but also see logistic growth).
- o Many responses of living beings to stimuli, including human perception, are logarithmic responses, which are the inverse of exponential responses; the loudness and frequency of sound are perceived logarithmically, even with very faint stimulus, within the limits of perception. This is the reason that exponentially increasing the brightness of visual stimuli is perceived by humans as a smooth (linear) increase, rather than an exponential increase. This has survival value. Generally it is important for the organisms to respond to stimuli in a wide range of levels, from very low levels, to very high levels, while the accuracy of the estimation of differences at high levels of stimulus is much less important for survival.
- o Computer technology

- o Processing power of computers. See also Moore's law and technological singularity (under exponential growth, there are no singularities. The singularity here is a metaphor.).
- o In computational complexity theory, computer algorithms of exponential complexity require an exponentially increasing amount of resources (e.g. time, computer memory) for only a constant increase in problem size. So for an algorithm of time complexity 2^x, if a problem of size x=10 requires 10 seconds to complete, then a problem of size x=11 will require 20 seconds, and x=12 will require 40 seconds. This kind of algorithm typically becomes unusable at very small problem sizes, often between 30 and 100 items (most computer algorithms need to be able to solve much larger problems, up to tens of thousands or even millions of items in reasonable times, something that would be physically impossible with an exponential algorithm). Also, the effects of Moore's Law do not help the situation much because doubling processor speed merely allows you to increase the problem size by one. E.g. if a slow processor can solve problems of size x in time t, then a processor twice as fast could only solve problems of size x+1 in the same time t. So exponentially complex algorithms are most often impractical, and the search for more efficient algorithms is one of the central goals of computer science.
- o Internet traffic growth.
- o Investment. The effect of compound interest over many years has a substantial effect on savings and a person's ability to retire.
- o Physics

- o Avalanche breakdown within a dielectric material. A free electron becomes sufficiently accelerated by an externally applied electrical field that it frees up additional electrons as it collides with atoms or molecules of the dielectric media. These secondary electrons also are accelerated, creating larger numbers of free electrons. The resulting exponential growth of electrons and ions may rapidly lead to complete dielectric breakdown of the material.
- o Nuclear chain reaction (the concept behind nuclear weapons). Each uranium nucleus that undergoes fission produces multiple neutrons, each of which can be absorbed by adjacent uranium atoms, causing them to fission in turn. If the probability of neutron absorption exceeds the probability of neutron escape (a function of the shape and mass of the uranium), $k > 0$ and so the production rate of neutrons and induced uranium fissions increases exponentially, in an uncontrolled reaction.
- o Multi-level marketing.

Exponential increases appear in each level of a starting member's downline as each subsequent member recruits more people.

THE QUANTUM POINT CONTACT (QPC)

A Quantum Point Contact (QPC) is a small point-like connection between two electrically-conducting regions. The typical size of such a constriction lies in the range of nano-to micrometre.

Fabrication

There are different ways of fabricating a QPC. It can be realised for instance in a break-junction by pulling apart a piece of conductor until it breaks. The breaking point forms the point contact. In a more controlled way, quantum point contacts are formed in 2-dimensional electron gases (2DEG), e.g. in GaAs/AlGaAs heterostructures. By applying a voltage to suitably-shaped gate electrodes, the electron gas can be locally depleted and many different types of conducting regions can be created in the plane of the 2DEG, among them quantum dots and quantum point contacts.

Another means of creating a point contact is by positioning an STM-tip close to the surface of a conductor.

Properties

Geometrically a quantum point contact is a constriction in the transverse direction which presents a resistance to the motion of electrons. Applying a voltage V across the point contact a current will flow, the size given by I = GV, where G is the conductance of the contact. This formula resembles Ohm's law for macroscopic resistors. However there is a fundamental difference here resulting from the small system size which requires a quantum mechanical point of view.

At low temperatures and voltages, electrons contributing to the current have a certain energy/momentum/wavelength called Fermi energy/momentum/wavelength. The transverse confinement in the quantum point contact results in a quantisation of the transverse motion much like in a waveguide. The electron wave can only pass through the constriction if it interferes constructively which for a given size of constriction only happens for a certain number of modes N. The current carried by such a quantum state is the product of the velocity times the electron density. These two quantities by themselves differ from one mode to the other, but their product is mode independent. As a consequence, each state contributes the same amount e2/h per spin direction to the total conductance

G = NGQ.

This is a fundamental result; the conductance does not take on arbitrary values but is quantised in multiples of the conductance quantum GQ = 2e2/h which is expressed through electron charge e and Planck constant h. The integer number N is determined by the width of the point contact and roughly equals the width divided by twice the electron wave length. As a function of the width (or gate voltage in the case of GaAs/AlGaAs heterostructure devices) of the point contact, the conductance shows a staircase behaviour as more and more modes (or channels) contribute to the electron transport. The step-height is given by GQ.

An external magnetic field applied to the quantum point contact lifts the spin degeneracy and leads to half-integer steps in the conductance. In addition, the number N of modes that contribute becomes smaller. For large magnetic fields N is independent of the width of the constriction, given by the theory of the quantum Hall effect.

An interesting feature, not yet fully understood, is a plateau at 0.7GQ, the so-called 0.7-structure.

Applications

Apart from studying fundamentals of charge transport in mesoscopic conductors, quantum point contacts can be used as extremely sensible charge detectors. Since the conductance through the contact strongly depends on the size of the constriction, any potential fluctuation (for instance, created by other electrons) in the vicinity will influence the current through the QPC. It is possible to detect single electrons with such a scheme. In view of quantum computation in solid-state systems, QPCs may be used as readout devices for the state of a qubit.

MOTORS OF SYNTHETIC MOLECULES

Synthetic molecular motors are nanoscale devices capable of rotation under energy input. Although the term "molecular motor" has traditionally referred to a naturally occurring protein that induces motion, some groups also use the term when referring to non-biological, non-peptide synthetic motors.

Many chemists are pursuing the synthesis of such molecular motors. The prospect of synthetic molecular motors was first raised by the nanotechnology pioneer Richard Feynman in 1959 in his classic talk There's Plenty of Room at the Bottom.

The basic requirements for a synthetic motor are repetitive 360° motion, the consumption of energy and unidirectional rotation. Two efforts in this direction were published in 1999 in the same issue of Nature. For the two reports below, it is unknown whether these molecules are capable of generating torque. It is expected that reports of more efforts in this field will increase, as understanding of chemistry and physics at the nanoscale improves.

Triptycene Motors

Molecular Motor (Kelly 1999). For clarity the aromatic rings of the triptycene moiety are omitted.In one molecular motor (Kelly, 1999) a three-bladed triptycene rotor connected to a rigid helicene scaffold is able to rotate 120° in a 5 step reaction sequence. The bond rotation barrier for the carbon carbon covalent bond connecting the two units and acting as the axle is 25 kcal/mol (105 kJ/mol).

In the first step of the sequence a molecule of phosgene is consumed converting the triptycene aniline group in

(1) into an isocyanate (2). The motor then picks up speed by thermally induced rotation which accounts for 10 kcal/mol (42 kJ/mol) (visualized in 3). This movement brings the isocyanate group in close proximity of the hydroxyl spacer mounted on the helicene part for a reaction to take place to the urethane (4).

This locks in the clockwise movement and thermal energy provides the second slow (80% conversion in 6 hours) rotation step (5). Note that the anticlockwise movement would move the two reactive groups away from each other. Finally the urethane bond is cleaved by sodium borohydride in ethanol to the original functional groups in the atropisomer (6) of the original molecule and the process can start again.

Helicene Motors

In 1999, the laboratory of Prof. Dr. Ben L. Feringa at the University of Groningen (The Netherlands) reported the creation of a monodirectional molecular rotor. Their 360° molecular motor system (Feringa, 1999) consists of a bis-helicene connected by an alkene double bond displaying axial chirality and having two stereocenters.

One cycle of unidirectional rotation takes 4 reaction steps. The first step is a low temperature endothermic photoisomerization of the trans (P, P) isomer 1 to the cis (M, M) 2 where P stands for the right handed helix and M for the left handed helix. In this process the two axial methyl groups are converted into two less sterically favorable equatorial methyl groups.

By increasing the temperature to 20 °C these methyl groups convert back exothermally to the (P, P) cis axial groups (3) in a helix inversion. Because the axial isomer is more stable than the equatorial isomer, reverse rotation is blocked. A second photoisomerization converts (P, P) cis 3 into (M, M) trans 4, again with accompanying formation of sterically unfavorable equatorial methyl groups. A thermal isomerization process at 60 °C closes the 360° cycle back to the axial positions.

A major hurdle to overcome is the long reaction time for complete rotation in these systems, which does not compare to rotation speeds displayed by motor proteins in biological systems. In the fastest system to date, with a fluorene lower half, the half life of the thermal helix inversion is 0.005 seconds. This fluorene compound is synthesized using the Barton-Kellogg reaction and rotation around the central double bond is believed to proceed much quicker due to a much higher energy of the ground state relative to the transition state in the thermal helix inversion. The previous record half-life was 3.2 minutes for the same compound with a methyl group and not the tert-butyl group.

The Feringa principle has been incorporated into a prototype nanocar. The car thus far synthesized has an helicene-derived engine with an oligo (phenylene ethynylene) chassis and four carborane wheels and is expected to be able to move on a solid surface with scanning tunneling microscopy monitoring, although so far this has not been observed. Interestingly the motor does not perform with fullerene wheels because they quench the photoexcited state of the motor moiety.

12

BENEFITS AND RISKS OF NANOTECHNOLOGY

In Simple word it can be said that all technologies are double-edged swords. Fundamentally new technological capabilities and benefits are accompanied by new risks, and new responsibilities for managing risks appropriately. The acceptance of these responsibilities is not optional. Dealing with the benefits and risks of nanotechnology openly and proactively, neither amplifying nor downplaying them, will be critical to the positive development of the field.

The first version of the Foresight Guidelines was developed during and after a workshop on Molecular Nanotechnology (MNT) Research Policy Guidelines sponsored by the Foresight Institute and the Institute for Molecular Manufacturing (IMM). The workshop was conducted over the February 19-21, 1999, weekend in Monterey, California. Participants included: James Bennett, Greg Burch, K. Eric Drexler, Neil Jacobstein, Tanya Jones, Ralph Merkle, Mark Miller, Ed Niehaus, Pat Parker, Christine Peterson, Glenn Reynolds, and Philippe Van Nedervelde. The Guidelines have been revised many times in the intervening years. The resulting Foresight Guidelines ("the Guidelines") include assumptions, principles, and some specific recommendations intended to provide a basis for the responsible development of nanotechnology.

Continued research and education are needed to create a shared understanding and sufficient knowledge base on the entire set of nanotechnology development and risk management issues that must be addressed. While discussion of guidelines can begin today, the scientific and technical community will continue to evolve its understanding of the issues. The Guidelines have already changed over time to reflect that dynamic understanding and specific feedback from a wider community.

The term nanotechnology refers to several distinct classes of technology, each with its own set of capabilities, potential applications, and risks. The specific terms used for these technologies vary over time; however, it is important to be clear about the fundamental distinctions between them.

ENGINEERING AND NANOSCALE SCIENCE

The nanoscale science and engineering conducted today has been defined as technology with a size range less than 100 nanometers or billionths of a meter. Using that definition only, most of chemistry would qualify.

The practitioners in the field would add that nanoscale science and engineering researches and exploits the unique properties of materials in the less than 100 nanometer size range. For example, depending on its specific configuration, carbon in that size range may exhibit extraordinary tensile strength greater than diamond, or act as an electrical conductor, insulator, or semiconductor. This branch of nanotechnology exists today with many research programs throughout the world, and many companies commercializing their applications. These companies may sell purified nanotubes made of tubular lattices of carbon, soccer ball like polyhedra made of 60 atoms of carbon called Buckyballs, dendritic polymers called dendrimers, or other particles in the less than 100 nanometer size range. The applications of these technologies are numerous and significant. They enable fundamentally new types of pharmaceuticals, electronic memory and semiconductor devices, sensors, renewable energy capture and storage systems, water purification devices, super strong fabrics and materials, security and military components, as well as antipollution devices. These applications are already beginning to emerge, and will gather momentum over the coming years.

The risks associated with passive compounds in the less than 100 nanometer size range concern their ability to be inhaled, absorbed through the skin, or to pass through biological compartment barriers such as the blood brain barrier. They thus pose a range of potential health and environmental risks that are associated with their potential toxicity or mutagencity in their interactions with biological systems. While the range of effects vary, most of the risks may be addressed by advanced industrial hygiene and environmental health practices and techniques that seek to characterize the specific risks, exposure patterns, and control methods and enforce them through a combination of practitioner education, industry self-regulation, monitoring and government regulation.

This is an important emerging field in the environmental and health sciences, since most of the existing legislation on environmental, safety, and health risks may cover particulates, but do not take the change in physical and biological properties at the nanoscale into account. It is reasonable to assume that passive nanoscale particle risks, although potentially serious if not addressed, will be characterized and addressed systematically under new versions or extensions to existing occupational, industrial hygiene, environmental, and medical regulations.

THE PRODUCTIVE NANOSYSTEMS

Productive nanosystems are currently a research oriented class of nanotechnology that will produce programmable, molecular-scale systems that make other useful nanostructured materials and devices. These systems may take over a decade to develop and mature.

They will be qualitatively different from nanomaterials, particularly regarding regulatory issues. These systems may be used as infrastructure for manufacturing, specifically the ability to build molecularly precise, inexpensive, three-dimensional products of arbitrary size.

The most straightforward infrastructure for manufacturing will be built with special purpose molecular fixtures and components that are analogous to macroscale factory components that produce devices that are inherently incapable of replication. These special purpose manufacturing systems will eventually be able to manufacture very large structures by scaling specific components and sub systems.

The simplest, most efficient, and safest approach to productive nanosystems is to make specialized nanoscale tools and put them together in factories big enough to make what is needed. People use simple tools to make more complex tools, from blacksmiths' tools to automated machinery. The convergent assembly architecture developed by Ralph Merkle (1997 Nanotechnology 8 18-22), describes how small parts can be put together to form larger parts, starting with nanoscale blocks and progressing up the hierarchy to macroscopic systems. The machines in this would work like the conveyor belts and assembly robots in a factory, doing similar jobs. If you pulled one of these machines out of the system, it would pose no risk, and be as inert as a light bulb pulled from its socket.

The eventual applications of these special purpose manufacturing systems include the ability to build almost any mechanical device cheaply, and in large quantity. This is why productive nanotechnology manufacturing capabilities will eventually do for our relationship to molecules and matter what the computer did for our relationship with bits and information. The computer enabled an ever expanding number of people to access billions of dollars worth of information. Productive nanotechnology will enable an ever expanding number of people to enjoy significant material wealth, based on carbon feedstock, which currently is in overabundant supply. It will also enable the technical infrastructure to address effectively many of our most pressing transportation, environmental, medical and global warming issues.

The primary risks of manufacturing enabled by productive nanosystems concern what is manufactured, not the manufacturing infrastructure itself. Special purpose manufacturing systems can be designed to be safe and reliable. They could be made to build a wide range of devices cheaply, in place, and on-demand. These products could include components for large scale buildings, computing, mass transit systems, energy storage, and spacecraft. On the other hand, they could also include tiny new security devices, and large quantities of inexpensive and super strong conventional weapon systems.

THE REPLICATORS

The concern about advanced forms of nanotechnology, after feasibility issues have been addressed, tends to center on the possibility of the technology getting out of control. Popular science fiction describes molecular robots that are autonomous self-replicating machines or autonomous replicators that evolve beyond human control, or can't stop reproducing. For the purpose of our Guidelines, an autonomous replicator is a specific kind of device that both (1) contains a set of materials processing and fabrication mechanisms sufficient to perform the operations necessary build devices like itself and (2) contains a set of design instructions and instruction-interpreting mechanisms sufficient to direct the operations necessary to build a device like itself. All other machines, lacking these exceptional properties, are not autonomous replicators. Productive nanotechnology enabled manufacturing of small or large products does not require any autonomous replicators, either in development or in application. It is also important to distinguish between the special purpose ability to manufacture many copies of a specific product, and the ability of the manufacturing infrastructure to replicate itself.

Nature employs its own biological form of molecular manufacturing to produce organisms. Productive nanosystems can be completely non-biological. They can be designed to function in a narrow and controlled range of physical conditions. Autonomous self-replicating assemblers are not necessary to achieve significant manufacturing capabilities. As Drexler and Phoenix indicated in their Safe Exponential Manufacturing paper (2004 Nanotechnology 15 869-872), developing manufacturing systems that use general purpose replicators able to extract their own energy sources is unnecessary.

It is important to note that molecular nanotechnology not specifically developed for manufacturing could be implemented as non autonomous replicating systems that have many layers of security controls and designed-in physical limitations. This class of system could potentially be used under controlled circumstances for nanomedicine, environmental monitoring, and specialized security applications. There are good reasons to believe that when designed and operated by responsible organizations with the appropriate quality control, these non autonomous systems could be made arbitrarily safe to operate. However, a determined and sophisticated group of terrorists or "non state entities" could potentially, with considerable difficulty, specifically engineer systems to become autonomous replicators able to proliferate in the natural environment, either as a nuisance, a specifically targeted weapon, or in the worst case, a weapon of mass destruction.

Both conventional nanoscale technology and manufacturing enabled by productive nanosystems can be implemented by responsible parties quite effectively without these risks, but as with other technologies, the risk of abuse must be considered seriously. Thus, in addition to the need for professional ethics and multiple layers of embedded industrial controls, there will also be a need for thoughtful regulation, monitoring, and potentially the development of "immune responses" to external threats.

Embedded Controls

There are many additional dimensions of safety controls that can be engineered into designs for replicators. Robert Freitas and Ralph Merkle (2004, Kinematic Self-Replicating Machines, Landes Bioscience, p. 152) have described a multidimensional space of 137 replicator design properties, characterized by structure, function, inputs, and outputs, and the percentage of its own components that a system can fabricate, extract, transport, inspect, warehouse, repair, control, or energize. Each of these design properties represents a potential control point for identification, embedded safety features, and deactivation. The authors support an earlier version of these Guidelines in their book, and agree with its recommendations against developing inherently unsafe replicator designs which permit: (1) surviving mutation or undergoing evolution, and (2) competing with or using biology as a raw materials resource.

Replicator designs may be made arbitrarily safe by employing layers of redundant embedded controls. For example, a broadcast architecture which transmits encrypted manufacturing instructions to a machine without an on-board instruction set. Another example is a vitamin architecture which designs in dependence on an exotic fuel or substance not available in the natural environment. While no technology is absolutely safe and free from the risk of abuse, we will refer to these systems as inherently safe replicator designs. Specifically, they are inherently safe compared to systems that are built without redundant layers of built-in safety controls. It is important to note that molecular manufacturing systems could be made considerably safer and more vulnerable to deactivation than natural bacteria or viruses which evolve rapidly, use biological materials as food, are readily available today, and are easier to hijack for nefarious purposes. Understanding the underlying properties of different forms of non autonomous replicating systems will make it much easier to engineer them with embedded safety and security features, as well as detect and control any attempts to abuse them.

Inherently safe designs may be needed for non autonomous replicators that can monitor the environment and respond rapidly to attacks or natural disasters. The development of nanotechnology is part of a general technological trend towards miniaturization that is proceeding globally at an accelerating pace. Bad actors who might try to abuse the technology could potentially be thwarted if replicator countermeasures or an "immune system" is deployed in advance. These systems would be analogous to those developed by nature for when organisms come under attack by viruses and bacteria, and anti-virus programs developed by humans to protect computing systems. Like our own very successful but imperfect immune system, this type of system could be compromised by sophisticated technologists and programmed to attack itself, as the body does in autoimmune diseases.

This does not mean immune defenses are useless. In fact, they have been selectively developed during millions of years of evolution. It does mean that we will need to build redundant layers of security into the design of these systems, and keep improving them rapidly. This may become a "predator-prey" cycle as competitors respond to the latest defense or attack mechanisms. More research and resources are required on the large space of design for safety in non autonomous replicator systems, design for security and rapid response in immune systems, and guidelines concerning their development and use.

Benefits and Risks

Policy related discussions of nanotechnology require consideration of the economic and environmental benefits of the different types of nanotechnology, as well as the potential problems. Some of the potential benefits of advanced forms of the technology include new lifesaving systems to address massive poverty and hunger, prevent and repair damage from diseases that are not responsive to today's antibiotics or anti-viral drugs, and provide rapid material response to natural disasters such as tsunamis, hurricanes, earthquakes, and volcanoes. Poverty, disease, and natural disasters kill thousands, in some cases millions annually, and the potential to ameliorate their effects significantly should not be relinquished lightly, particularly by those least affected.

The Foresight Nanotechnology Challenges address critical needs that could be met by developing a range of near and long term nanotechnology solutions. They include:

(1) meeting global energy needs through more efficient generation, storage and distribution,

(2) providing abundant clean water through improved water purification and filtration,

(3) increasing health and longevity of human life through medical diagnostics, drug delivery and customized therapy,

(4) maximizing the productivity of agriculture through precision farming, targeted pest management and the creation of high yield crops,

(5) making powerful information technology available everywhere through reduced cost and higher performance of memory, networks, processors and components, and *6) enabling the development of space resources through improved fuels, as well as smart materials and environments.

Meeting any one of these challenges with nanotechnology would have a transformative effect on our quality of life. Thus, the need for new controls should not prevent the responsible development of the field.

There has been considerable confusion about the sources of risk in the development and future deployment of nanotechnology. Most of the risk assessment of current nanoscale technology concerns compatibility of nanomaterials with humans and the environment. Relevant ecological and public health principles must be utilized in the development of any technology, particularly one as fundamental and broad as nanotechnology. Nanoparticles may be absorbed or inhaled if the proper industrial hygiene precautions are not utilized. They may also pass readily through body compartments such as the blood brain barrier. Manufactured diamondoid products may not break down easily in the natural environment. Consumers may not initially have means readily available to recycle them. Thus, total "product lifecycle" considerations about health and environmental effects should be taken into consideration as industry develops new manufacturing techniques based on productive nanosystems.

Since some controls on the different classes of nanotechnology will eventually be put into place, it makes sense for them to be as well informed as possible. Rather than have external controls imposed upon an R&D community that is not addressing potential risks openly, the developing nanotechnology R&D community should adopt appropriate self-imposed controls proactively. They can also participate in policy discussions on external controls that may be formulated in light of current knowledge and the evolving state of the art. The quantity and quality of these additional controls will depend to some extent on the success of voluntary controls.

SUCCESSFUL PRECEDENT

The NIH Guidelines on Recombinant DNA technology are an example of self-regulation taken by the biotechnology community over 30 years ago. While the kind of artificial molecular machines of primary interest for nanotechnology are expected to be very different from the kind of biological systems covered by the NIH

Guidelines (just as a 747 is very different from a sparrow, even though both fly), the NIH Guidelines illustrate that advance preparations are possible and can be effective.

Those guidelines were so well accepted that the privately funded research community has continued to submit research protocols for juried review, in spite of the fact that it was optional for them to do so. In addition, although the NIH Guidelines have been progressively relaxed since they were first released, they did achieve their intended effect.

IMPROVING OPPORTUNITIES REDUCING RISKS

Industry and government should have the maximum opportunity to develop and commercialize a manufacturing industry based on productive nanosystems designed for safety and reliability. In addition, nanotechnology should be developed in ways that make it possible to distribute the substantial benefits of the technology to the majority of humanity currently desperate to achieve material wealth at any environmental or security cost. Manufacturing based on productive nanosystems will eventually be capable of producing widespread material abundance with significantly less environmental impact than technologies in common use today. Providing technical abundance alone cannot make a country wealthy and secure.

This also requires education, and social arrangements that include rule of law, and other features of civil society. However, technological abundance can alleviate many of the conflicts that stem primarily from rivalry over resources. Reducing this specific cause of conflict could make the world considerably more secure than it is today. In addition, the release from bare economic subsistence could enable billions of people to take advantage of the emerging global classroom enabled by the World Wide

Web. This education effect could compound the positive security benefits of nanotechnology.

Given the accelerating world wide research on various types of nanotechnology it is highly probable that advanced capabilities will eventually be developed, and thus, it is important for society to consider proactively the range of controls that should be in place to minimize risks and maximize economic benefits. Unlike the proliferation of nuclear technology, which can be partially controlled by limiting access to fissionable materials, once the technology matures, it could utilize carbon, which is readily available. Further, there is no guarantee that limiting research and development on these systems in open and democratic countries would effectively slow their development and deployment elsewhere. China, for example, has an active nanotechnology research program, and it is training large numbers of qualified scientists and engineers in the relevant disciplines for developing advanced forms of the technology. The Guidelines take the position that the safest and most responsible path is to enact reasonable technology-specific controls at the practitioner, industry, and governmental levels, and simultaneously develop the monitoring and counter measures to control potentially rogue or offensive use of the technology.

REGULATION AND TREATIES

Effective means of restricting the misuse of molecular nanotechnology in the international arena will need to be developed. If weaponized versions of MNT are developed, they may not fall under existing arms-control treaties. Adding productive nanosystems designed for manufacturing to the list of technologies covered in Chemical, Biological and Nuclear Weapons treaties could be inappropriate because MNT is not a weapon, but a productive technology with broad applications. It is more similar to chemical technology than to chemical weapons, and more similar to biotechnology than to biological weapons.

Adding particular weapons related applications of MNT to the list of technologies covered in Chemical, Biological and Nuclear Weapons treaties may be appropriate in certain cases. It should be remembered, however, that the capabilities of productive nanosystems will be extensions of general manufacturing technology. The military applications of MNT will include the manufacture of high performance aerospace vehicles and precision munitions at low cost. The high value and dual use of MNT for civilian and defense purposes will require making distinctions between the enabling technology and its specific applications, balancing health and economic benefits against security concerns. Since nanotechnology research is now global, the security challenges will be present, with or without the ability to capture the wide variety of benefits.

Overly restrictive treaties or regulatory regimes applied to core MNT technologies could lead to the unintended consequence that only the rule-following nations would be at a competitive disadvantage technologically, economically, and militarily. While most nations, companies, and individuals are likely to adhere to reasonable safety restrictions, guidelines that are viewed as too restrictive will simply be ignored, paradoxically increasing risk. In addition, not all guidelines and laws will be followed, and enforcement is rarely perfect. Non-state actors could become quite significant, particularly when the relevant knowledge and raw materials are available globally. They may well not be signatories to any agreement. While a 100% effective ban could, in theory, avoid the potential risks of certain forms of molecular nanotechnology, a 99.99% effective ban could result in development and deployment by the 0.01% that evaded and ignored the ban. For example, the international Biological Weapons Treaty was being violated on a massive scale even before the ink was dry.

On the other hand, international cooperation on restricting the proliferation and use of atomic weapons has been partially effective in limiting their development and use over the past several decades. In that case, however, the raw materials were not widely available. There are reasonable arguments on both sides of the treaty question. It is wise to avoid an unnecessary nanotechnology arms race, particularly when manufacturing enabled by productive nanosystems could be used to greatly reduce competition for material resources and mutually improve quality of life for the competitors. Treaties may become easier to verify, at least when the verifiers are more advanced than the verified. The "trust, but verify" security concept will increase in importance in a potentially dangerous, but increasingly transparent world. In fact, nanotechnology based sensors are likely to increase multifold the options for transparency.

The Guideline participants as a group have not endorsed any specific means to address MNT security concerns through treaty arrangements. However, as nanotechnology capabilities increase, governmental, NGO, industry cooperative arrangements, and self-funding enforcement mechanisms will need to be implemented. Self policing NGOs and industry groups will play an important role in this, but the public is likely to want some oversight provided by elected representatives. This may reflect distinctions between addressing potential voluntary and localized risks vs. involuntary and more widespread risks, as well as differences in the distribution of benefits. In addition, monitoring defense applications will have to be handled primarily by government agencies. There will be no easy answers in this arena, only a set of dynamic trade-offs that will have to be incorporated into flexible organizations and policies that can survive the vagaries of the real world of politics, economics, and accelerating technology.

The international community of nations, industry, and nongovernmental organizations will need to develop effective means of enforcing nanotechnology guidelines and regulations, and responding to misuse. Such means should not restrict the development of peaceful applications of the technology or defensive measures by responsible members of the community. Today there are obvious difficulties in achieving consensus on the definition of "responsible members" and enforcing international agreements on chemical and biological weapons is known to be problematic. However, given the importance of making progress in this area, further research, collaboration, and innovation is encouraged.

THE EDUCATION AND ENFORCEMENT

The safe development and use of nanotechnology depends, in part, on the good judgment and ethical behavior of the researchers carrying out this work. This is an imperfect, but important first line of defense. The more this is recognized as critically important, the more effective the vast majority of researchers are likely to be in actively preventing unsafe designs or uses of nanotechnology, and in insuring that manufacturing systems have built-in safeguards. The natural and responsible path for the development of productive nanotechnology based manufacturing makes use of no autonomous replicators. However, defense against potential rogue elements who might seek to abuse replicators is a problem not unlike the challenge of controlling the developers of viruses on the Internet. In both cases, a combination of moral and technical education, active industry and government cooperation, inherently safe system designs, legal frameworks, and R&D on secure immune systems for defense may be the best solutions available.

Nanotechnology policy will have to balance risks with benefits, and distinguish between different classes of risks. Molecular manufacturing and nanotechnology are not one technology, but rather a spectrum of technologies, with radically different risk profiles. A substantial R&D program is needed to clarify the nature, magnitude and likelihood of the potential risks, as well as the options available for dealing with them effectively. For example, toxicology analyses relating to nanomaterials have already been identified as an early priority. Nanomaterial safety is a matter that is distinct from the key risks of productive nanotechnology based manufacturing, but both require good industrial hygiene practices.

There are significant risks associated with failing to address the increasingly costly economic, political, environmental, energy, and security problems that the development of productive nanosystems could help resolve. Likewise, there are real costs to restrictive policies that limit nanotechnology innovation by responsible actors and allow rogue entities to move ahead. The Guidelines were not intended to cover every risk or potential abuse of the technology. People may abuse automobile technology, and society has responded by making cars safer to operate, holding drivers accountable for their actions through laws that are enforced, and requiring drivers to pay for automobile insurance. Likewise, industry and governments are held responsible for their use of technologies that have widespread impact.

The Guidelines are intended to cover most of the risks associated with normal development and use of the technology, and to mitigate, as much as possible, the risks associated with potential abuse of the technology. However, most guidelines and embedded control regimes can be circumvented by a sophisticated and determined adversary. Defending against this kind of threat may require an active monitoring and immune system for detection and deactivation.

Informed policy decisions could accelerate the safe development of peaceful and environmentally responsible uses of nanotechnology. This includes capturing the opportunity to develop powerful new approaches to medicine, and energy efficient, zero emission manufacturing and transportation technologies. Informed and thoughtful policies will also increase our security by enabling new types of immune systems to detect and respond to deliberate abuse.

The field of nanotechnology is very broad, like computing, and thus its eventual regulation spans human health and safety (NIH), environmental protection (EPA), and eventually weapon systems (DoD, DHS, CIA). It is important that appropriate distinctions be made between different classes of nanotechnology in the development of regulations, and that effective interagency coordination is done to promote and enforce consensus, and build on existing regulatory standards and monitoring where appropriate.

The self assessment scorecards are based on the notion that the people, organizations, and governments that work in the nanotechnology field should develop and utilize professional guidelines and practices. These guidelines and practices should be grounded in science and technology principles, and knowledge of the interacting environmental, security, ethical, and economic issues relevant to the development of the field.

This is based on the notion that professional ethics, "soft law", and cultural norms regarding good practice are at least as effective as "hard law" in preventing unsafe practices, and in helping to ensure that unsafe practices are noticed and acted upon. The use of "soft laws" is a first line of defense, and is not meant to suggest that "hard laws" for safety and health are not useful, and at times appropriate.

Any regulation adopted by researchers, industry or government should provide specific, clear guidelines. Regulators should have specific and clear mandates, providing efficient and fair methods for identifying different classes of hazards and for carrying out inspection and enforcement.

There is great value in seeking the least-restrictive necessary legal environment to ensure the safe and secure development of each specific type of nanotechnology. It is important to recognize in this context that some types of nanotechnology will eventually provide the best solutions available for remedying the existing environmental and public health damage resulting from our current, distinctly suboptimal, technology base.

Scorecard 1: Nanotechnology Professional Guidelines

Self Scoring: 0-5,

0 = no compliance,

5 = high compliance

Best Score in this section = 40

1. Nanotechnology developers adopt professional guidelines and ethical practices relevant to the responsible development of both near term and advanced nanotechnology.

2. Nanotechnologists attempt to consider proactively and systematically the environmental and health consequences of their specific technologies. Practitioners recognize that the scope and magnitude of potential problems are reduced to the extent that they consider the range of possible negative consequences, and plan to prevent them, or at least minimize their effects through embedded and redundant control systems.
3. Nanotechnology research and development is conducted with due regard for the principles of environmental science and standard practices of public health, with the understanding that significant changes in physical, chemical, and physiological properties may occur when macroscale materials are developed and utilized on the nanoscale.
4. Nanotechnology products are conceived and developed using total product lifecycle analysis.
5. Productive nanosystem based manufacturing makes use of inherently safe system designs requiring no autonomous replication.
6. When controversy exists concerning the theoretical feasibility or implementation timing of advanced nanotechnologies, such as specialized manufacturing components or scaling techniques, researchers address and clarify the issues rapidly, and attempt to resolve any controversy openly.
7. Any developers who consider the design or development of non autonomous replicators for specific R&D purposes should first explore the potential benefits and risks of alternative approaches actively, in a balanced and rigorous manner.

8. Any use of potentially autonomous replicators is avoided in manufacturing, and only utilized in R&D after institutional review and approval of extensive and redundant control systems.

Scorecard 2: Nanotechnology Industry Guidelines

Self Scoring: 0-5, 0 = no compliance,

5 = high compliance

Best Score in this section = 40

1. Industry self-regulation is practiced proactively, and tailored to the specific risk profile of the nanotechnology under development. Specifically, studies to assess the consequences of new nanotechnology processes, materials, and tools are scoped appropriately, advanced as rapidly as possible, and encompass both benefits and risks with rigor.
2. Manufacturing systems are described and classified according to their specific characteristics, particularly with respect to autonomy and safety control systems.
3. When molecular manufacturing systems are implemented, they use inherently safe system designs with no autonomous replicators.
4. Any molecular manufacturing device designs specifically limit unplanned distribution and provide traceability and audit trails.
5. Encrypted molecular manufacturing device instruction sets are utilized to discourage misuse.
6. Autonomous replicators for R&D are avoided through the use of inherently safe system designs that selectively utilize non autonomous system characteristics, and layers of redundant controls.

7. Replication systems used for R&D are designed to be incapable of autonomous replication in any natural environment. They have multiple system requirements (e.g., for externally supplied information, interventions, environmental conditions, materials, components, or exotic energy sources) that are available only where deliberately provided to enable operation of the machine.
8. Replicator R&D focused on detecting and responding to potential technology threats utilizes redundant embedded safety controls such as time limited operations, encrypted external controls to override internal operations, and anti-mutation protections. For example, the information that specifies their construction is stored and copied in encoded form, and the encoding is such that any error in copying randomizes and thus destroys the decoded information. These systems are continuously improved for security.

Scorecard 3 Government Policy Guidelines

Self Scoring: 0-5, 0 = no compliance,

5 = high compliance

Best Score in this section = 55

1. Regulatory studies and controls distinguish the wide variety of nanotechnologies, and recognize that their different risk profiles require different regulatory policies.
2. Regulations and consensus standards promulgated by researchers, industry, or government provide specific and clear guidelines, and encourage the use of inherently safe system designs for manufacturing and R&D.

3. The government has designated a division within a regulatory entity or a new agency with sufficient resources to ensure nanotechnology standards enforcement and coordination across agencies. It is building upon and when necessary augmenting existing regulatory structure and institutions (eg. for health and safety, environment, defense, and intelligence). Regulators have specific responsibilities and authorities for identifying different classes of hazards, providing development approvals when necessary, and for carrying out inspection and enforcement.

4. Economic incentives are provided for responsible innovation through discounts on liability insurance policies, access to royalties, or consortia membership for molecular manufacturing and development organizations that certify Guidelines compliance. Willingness to provide self-regulation and open access for third party inspection that safeguards proprietary technology is a condition to utilize advanced forms of molecular nanotechnology.

5. Access to special purpose productive nanosystems enabled manufacturing with inherently safe designs and no autonomous replication is unrestricted, unless the special purpose capabilities pose a specific risk.

6. Initiatives are in place to encourage the international community and non-governmental organizations to restrict the deliberate misuse of molecular nanotechnology by improving verification, monitoring, and detection techniques, and making the detection and enforcement of misconduct increasingly probable. Such means should not restrict the development of nanoscale

materials, special purpose manufacturing systems, or non autonomous defensive measures utilizing inherently safe designs.

7. Accidental or willful misuse of nanotechnology is further constrained by legal liability and, where appropriate, subject to criminal investigation and prosecution. This should also pertain to those that enable and collaborate on the misuse of the technology.
8. Eventual distribution of advanced molecular nanotechnology capability is restricted, whenever possible, to responsible actors that have agreed to practice these Guidelines, and permit verification. No such restriction need apply to special-purpose, molecular machine systems with no autonomous replicators and inherently safe designs, or to the end products of molecular manufacturing that satisfy the Guidelines.
9. Governments, companies, and individuals who fail to follow reasonable principles and guidelines for development and dissemination of MNT are placed at a substantial competitive disadvantage with respect to access to companies, collaborative NGO organizations, R&D funding, plans, designs, software, hardware, and cooperative market relationships.
10. Incentives are in place to encourage industry, government, and NGO developers to collaborate on continuous improvement and use of best practices in nanotechnology and risk management, including the theory, mechanisms, and experimental designs for increasingly safe manufacturing, as well as effective monitoring and control systems.

11. Regulatory entities sponsor research on increasing the accuracy and fidelity of environmental and health models used for nanotechnology risk assessment and management, as well as the theory, mechanisms, and experimental designs for built-in safeguards and defensive nanodevice immune systems.

THE BACKDROP

The idea of guidelines for the safe development of nanotechnology has been discussed within the Foresight community for over a decade. It is inevitable that any guidelines put forth will be further discussed and perhaps substantively changed; but the dialog on specific proposals must begin somewhere. This latest version of the Foresight Guidelines represents another step in an ongoing discussion. In spite of the diversity of briefing materials and views represented at the initial Monterey workshop in February of 1999, the participants managed to discuss the technical and policy issues with both intensity and civility. While any one participant might have preferred more or less emphasis on a particular issue, the group was able to converge on a common set of draft guidelines for the development of nanotechnology.

The group agreed to review the Guidelines among themselves, discuss them in wider Foresight meetings during 1999, and then release them on the internet for review by the larger community. The goal was to get the Guidelines endorsed and adopted by organizations sponsoring nanotechnology research and development projects, and to inspire effective self-regulation wherever necessary and possible. Explicit discussion of long term risks and potential regulation of industry made acceptance difficult. Another goal of the Workshop members was to educate nanotechnology researchers about the potential benefits and risks of the technology. The long-term goal was to eventually produce a dialog and set of Guidelines that would be useful to policy makers, the public, and the MNT research and development community. We believe that this has happened.

The Foresight Guidelines are intended as a living document, subject to modification and revision. Early drafts have been reviewed and revised several times since the Monterey workshop, including during Foresight/IMM sponsored discussions led by Neil Jacobstein in May and November of 1999. They were also provided in the attachments to Ralph Merkle's June 1999 Congressional testimony on MNT, and referenced in Neil Jacobstein's presentation on Nanotechnology and Molecular Manufacturing: Opportunities and Risks at Stanford University's Colloquium for Doug Engelbart in January of 2000. The Workshop participants debated whether the Guidelines were sufficiently developed for widespread publication, when Bill Joy's article: "Why the Future Doesn't Need Us" was published in the April 2000 issue of Wired Magazine. This article raised public awareness of the potential dangers of self-replicating technologies, including nanotechnology.

Since that time, the Guidelines were reviewed critically by Robert Freitas, and revised by Ralph Merkle and Neil Jacobstein. Version 3.6 of the Guidelines was discussed in a May 2000 Foresight workshop session led by Neil Jacobstein. Bill Joy was invited to participate in this discussion. He made several constructive suggestions, including one that outlined a guideline on closing the economic incentives loop via an insurance policy requirement for developers. Jacobstein incorporated the feedback from this and subsequent discussions into version 3.7 of the Guidelines, and they were then published for open review on the web.

Neil Jacobstein rewrote the Guidelines as a form of self assessment scorecards for version 4.0, based on the observation that this kind of self assessment is becoming a standard part of quality and six sigma programs in industry and government. He combined and added some new guidelines, including a guideline based on a paper by Eric Drexler and Chris Phoenix in the Journal of Nanotechnology on "Safe Exponential Manufacturing". This paper made the case for nanotechnology enabled manufacturing using a hierarchy of machine tools, without the need for general purpose self-replicating assemblers. Glenn Reynolds edited the draft and provided an analysis by his law students on current treaties and the fact that weaponized MNT might not be covered by them. Eric Drexler also reviewed the draft and made additional editorial suggestions.

Jacobstein presented version 4 of the Guidelines at the Foresight Advanced Molecular Nanotechnology Conference in Washington DC in September of 2004, and discussed revised versions at a National Academy of Sciences workshop on the Feasibility of Molecular Manufacturing held in February of 2005, and an Aspen Institute Seminar on Future Perspectives in July 2005. He wrote version 5.0 with feedback from these meetings and specific comments on the Guidelines or the risks they address from Robert Freitas, Marc Gubrud, Eric Drexler, Ray Kurzweil, Bill Joy, Martine Rothblatt, Larry Millstein, Max Moore, Doug Mulhall, Christine Peterson, David Forrest, Chris Phoenix, and Glenn Reynolds. Robert Freitas and David Forrest provided detailed reviews of this draft. Jacobstein produced and presented version 5 of the Guidelines for the October 2005 Foresight Advanced Molecular Nanotechnology Conference in San Francisco. He produced a substantially revised version 6.0 based on feedback from John Bashinski, Christine Peterson, David

Forrest and others. David Forrest referenced the Guidelines in his presentation to the Roundtable Discussion on Nanotechnology Regulation of the Senate Subcommittee on Environment and Public Works, April 6, 2006. The Guidelines represent a complex set of tradeoffs between competing concerns, and it must satisfy the needs of more than one special interest community. Thus, it is likely that few of its participants agree with all of it, though most would acknowledge its value. We acknowledge their value in improving the Guidelines.

Eventually, the Guidelines need to become sufficiently specific that they can form the basis for a legally enforceable framework within which nanotechnology development can be safely pursued. Future versions of the Guidelines or legislation inspired by them might eventually be enforced via a variety of means, possibly including lab certifications, randomized open inspections, professional society guidelines and peer pressure, insurance requirements and policies, stiff legal and economic penalties for violations, and other sanctions. Enforcement will be inherently imperfect, but the deterrent effect of unpredictable inspection, combined with predictable and swift consequences for violations, may prove preferable to the available alternatives.

Care must be taken that future revisions of the Guidelines do not become so restrictive that they simply drive nanotechnology research and development underground. This could expose compliant countries to the increased risks associated with decreased technical, economic, and military capabilities. It would also sacrifice the many significant economic, environmental, and medical benefits of nanotechnology that counteract serious and certain risks that society now faces in industrialized countries, and particularly in the developing world.

13

NANOTECHNOLOGY AND APPLICATIONS OF ENERGETIC FORMS

Over the past few decades, the fields of science and engineering have been seeking to develop new and improved upon types of energy technologies that have the capability of improving life all over the world. In order to make the next leap forward from the current generation of technology, scientists and engineers have been developing a new field of science called Nanotechnology. Nanotechnology is simply any technology that contains components smaller than 100 nanometers. For scale, a single virus particle is about 100 nanometers in width.

An important subfield of Nanotechnology related to energy is Nanofabrication. Nanofabrication is the process of designing and creating devices on the nanoscale. Creating devices smaller than 100 nanometers opens many doors for the development of new ways to capture, store, and transfer energy.

The inherent level of control that Nanofabrication could give scientists and engineers would be critical in providing the capability of solving many of the problems that the world is facing today related to the current generation of energy technologies.

People in the fields of science and engineering have already begun developing ways of utilizing Nanotechnology for the development of consumer products. Benefits already observed from the design of these products are an increased efficiency of lighting and heating, increased electrical storage capacity, and a decrease in the amount of pollution from the use of energy. Benefits such as these make the investment of capital in the research and development of Nanotechnology a top priority.

CONSUMER PRODUCTS

Recently, previously established and entirely new companies such as BetaBatt, Inc. and Oxane Materials are focusing on Nanomaterials as a way to develop and improve upon older methods for the capture, transfer, and storage of energy for the development of consumer products.

ConsERV, a product developed by the Dais Analytic Corporation, uses nanoscale polymer membranes to increase the efficiency of heating and cooling systems and has already proven to be a lucrative design. The polymer membrane was specifically configured for this application by selectively engineering the size of the pores in the membrane to prevent air from passing, while allowing moisture to pass through the membrane. Polymer membranes can be designed to selectively allow particles of one size and shape to pass through while preventing others of different dimensions. This makes for a powerful tool that can be used in consumer products from biological weapons protection to industrial chemical separations.

A New York based company called Applied NanoWorks, Inc. has been developing a consumer product that utilizes LED technology to generate light. Light Emitting Diodes or LEDs, use only about 10% of the energy that a typical incandescent or fluorescent light bulb use and typically lasts much longer, which makes them a viable alternative to traditional light bulbs. While LEDs have been around for decades, this company and others like it have been developing a special variant of LED called the white LED. White LEDs consist of semi-conducting organic layers that are only about 100 nanometers in distance from each other and are placed between two electrodes, which create an anode, and a cathode. When voltage is applied to the system, light is generated when electricity passes through the two organic layers. This is called electroluminescence. The semiconductor properties of the organic layers are what allow for the minimal amount of energy necessary to generate light. In traditional light bulbs, a metal filament is used to generate light when electricity is run through the filament. Using metal generates a great deal of heat and therefore lowers efficiency.

Research for longer lasting batteries has been an ongoing process for years. Researchers have now begun to utilize Nanotechnology for battery technology. mPhase Technologies in conglomeration with Rutgers University and Bell Laboratories have utilized nanomaterials to alter the Wetting Behavior of the surface where the liquid in the battery lies to spread the liquid droplets over a greater area on the surface and therefore have greater control over the movement of the droplets. This gives more control to the designer of the battery. This control prevents reactions in the battery by separating the electrolytic liquid from the anode and the cathode when the battery is not in use and joining them when the battery is in need of use.

Economic Benefits

The relatively recent shift toward using Nanotechnology with respect to the capture, transfer, and storage of energy has and will continue to have many positive economic impacts on society. The control of materials that Nanotechnology offers to scientists and engineers of consumer products is one of the most important aspects of Nanotechnology. This allows for an improved efficiency of products across the board.

A major issue with current energy generation is the loss of efficiency from the generation of heat as a by-product of the process. A common example of this is the heat generated by the internal combustion engine. The internal combustion engine losses about 36% of the energy from gasoline as heat and an improvement of this alone could have a significant economic impact. However, improving the internal combustion engine in this respect has proven to be extremely difficult without sacrificing performance. Improving the efficiency of fuel cells through the use of Nanotechnology appears to be more plausible by using molecularly tailored catalysts, polymer membranes, and improved fuel storage.

In order for a fuel cell to operate, particularly of the hydrogen variant, a noble-metal catalyst (usually platinum, which is very expensive) is needed to separate the electrons from the protons of the hydrogen atoms. However, catalysts of this type are extremely sensitive to carbon monoxide reactions. In order to combat this, alcohols or hydrocarbons compounds are used to lower the carbon monoxide concentration in the system. This adds an additional cost to the device. Using nanotechnology, catalysts can be designed through nanofabrication that are much more resistant to carbon monoxide reactions, which improves the efficiency of the process and may be designed with cheaper materials to additionally lower costs.

Fuel cells that are currently designed for transportation need rapid start-up periods for the practicality of consumer use. This process puts a lot of strain on the traditional polymer electrolyte membranes, which decreases the life of the membrane requiring frequent replacement. Using Nanotechnology, engineers have the ability to create a much more durable polymer membrane, which addresses this problem. Nanoscale polymer membranes are also much more efficient in ionic conductivity. This improves the efficiency of the system and decreases the time between replacements, which lowers costs.

Another problem with contemporary fuel cells is the storage of the fuel. In the case of hydrogen fuel cells, storing the hydrogen in gaseous rather than liquid form improves the efficiency by 5%. However, the materials that we currently have available to us significantly limit fuel storage due to low stress tolerance and costs. Scientists have come up with an answer to this by using a nanoporous styrene material (which is a relatively inexpensive material) that when super-cooled to around-196oC, naturally holds on to hydrogen atoms and when heated again releases the hydrogen for use.

CAPACITORS

For decades, scientists and engineers have been attempting to make computers smaller and more efficient. A crucial component of computers are capacitors. A capacitor is a device that is made of a pair of electrodes separated by an insulator that each stores an opposite charge. A capacitor stores a charge when it is removed from the circuit that it is connected to; the charge is released when it is replaced back into the circuit. Capacitors have an advantage over batteries in that they release their charge much more quickly than a battery.

Traditional or foil capacitors are composed of thin metal conducting plates separated by an electrical insulator, which are then stacked or rolled and placed in a casing. The problem with a traditional capacitor such as this is that they limit how small an engineer can design a computer. Scientists and engineers have since turned to Nanotechnology for a solution to the problem.

Using nanotechnology, researchers developed what they call "ultracapacitors." An ultracapacitor is a general term that describes a capacitor that contains nanocomponents. Ultracapacitors are being researched heavily because of their high density interior, compact size, reliability, and high capacitance. This decrease in size makes it increasingly possible to develop much smaller circuits and computers. Ultracapacitors also have the capability to supplement batteries in hybrid vehicles by providing a large amount of energy during peak acceleration and allowing the battery to supply energy over longer periods of time, such as during a constant driving speed. This could decrease the size and weight of the large batteries needed in hybrid vehicles as well as take additional stress off the battery. However, as of now, the combination of ultracapacitors and a battery is not cost affective due to the need of additional DC/DC electronics to coordinate the two.

Nanoporous carbon aerogel is one type of material that is being utilized for the design of ultracapacitors. These aerogels have a very large interior surface area and can have its properties altered by changing the pore diameter and distribution along with adding nanosized alkali metals to alter its conductivity.

Carbon nanotubes are another possible material for use in an ultracapacitor. Carbon nanotubes are created by vaporizing carbon and allowing it to condense on a surface. When the carbon condenses, it forms a nanosized tube composed of carbon atoms. This tube has a high surface area, which increases the amount of charge that can be stored. The low reliability and high cost of using carbon nanotubes for ultracapacitors is currently an issue of research.

CAPACITANCE THEORY

Understanding the concept of capacitance can be helpful in understanding why Nanotechnology is such a powerful tool for the design of higher energy storing capacitors. A capacitor's capacitance (C) or amount of energy stored is equal to the amount of charge (Q) stored on each plate divided by the voltage (V) between the plates. Another representation of capacitance is that capacitance (C) is approximately equal to the permittivity (e) of the dielectric times the area (A) of the plates divided by the distance (d) between them. Therefore, capacitance is proportional to the surface area of the conducting plate and inversely proportional to the distance between the plates.

Using carbon nanotubes as an example, a property of carbon nanotubes is that they have a very high surface area to store a charge. Using the above proportionality that capacitance (C) is proportional to the surface area (A) of the conducting plate; it becomes obvious that using nanoscaled materials with high surface area would be great for increasing capacitance. The other proportionality described above is that capacitance (C) is inversely proportional to the distance (d) between the plates. Using nanoscaled plates such as carbon nanotubes with nanofabrication techniques, gives the capability of decreasing the space between plates which again increases capacitance.

Center on Nanotechnology and Society

The Center on Nanotechnology and Society (Nano & Society) is an affiliate of Illinois Institute of Technology

(IIT) and is housed at IIT's Chicago-Kent College of Law. Nano & Society is an affiliate of the Institute on Biotechnology and the Human Future, also based at Chicago-Kent College of Law.

Purpose

The Center on Nanotechnology and Society (Nano & Society) was created to catalyze informed interdisciplinary research, education, and dialogue on the ethical, legal, economic, policy, and broader societal implications of nanoscale science and technology. As an affiliate of the Institute on Biotechnology and the Human Future, Nano & Society approaches nanotechnology with a special focus on the human condition. Nano & Society brings the foremost scholars and researchers in nano science, law, ethics, medicine, and the social sciences together with leaders in business and industry. Nano & Society also employs a number of vehicles in its efforts to focus a national conversation on nanotechnology and human dignity, including its website, the Chicago Nano Forum, national conferences, and NELSI Global (a public posicy document archive).

Online Offerings

Nano & Society conducts research on NELSI--Nanotechnology's ethical, legal, and social implications. The Center features various initiatvies: NELSI Global is a web-based global public policy document archive focused on nanotechnology's ethical, legal and societal implications (NELSI). This initiative is accessible through Nano & Society's website. NELSI Global serves as a unified clearinghouse, enabling users to identify and access key public policy documents addressing the NELSI agenda. It includes international and U.S.-specific regulations, legislation, case law, congressional testimony, and key governmental reports.

A website that addresses: law, policy, and regulation; environment, health and safety; business; applications; and Publications

Publications, Conferences, and Events

The Center on Nanotechnology and Society uses multiple communications vehicles to educate and disseminate information on NELSI. These mechanisms include:

Nano and Society: The Center on Nanotechnology and Society publishes a monthly e-newsletter, entitled Nano & Society. which features opinion pieces from guest writers on nanotechnology's potential and potential impact, as well as public attitudes toward and education about nanotechnology; and relevant news updates on nanotechnology.

Nanologues: The Center on Nanotechnology and Society publishes a print series of booklets on NELSI entitled Nanologues.

Conferences and Events: Nano & Society hosts various national events to advance the dialogue on nanotechnology and the human condition, including:

Chicago Nano Forum: In conjunction with its partners in the educational, business, environmental, legal, and other arenas, Nano & Society has developed the Chicago Nano Forum. The Forum encourages public dialogue among nano experts in science, business, social sciences, ethics, and law. This series of events is webcast on the Center's website.

1st Annual Conference on Nanopolicy and the Human Future: This Nano & Society conference provided members of Congress and their staff, researchers, scientists, and others interested in and involved with the nano sector with the latest nanotechnology developments in the ethical, legal and social arena.

Center For Responsible Nanotechnology

The Center for Responsible Nanotechnology (CRN), founded in December, 2002, is a non-profit research and advocacy organization with a focus on molecular manufacturing and its possible effects, both positive and negative.

CRN provides information to journalists, business leaders, policymakers and the general public about the environmental, humanitarian, economic, military, political, social, medical, and ethical implications of advanced nanotechnology.

CRN is an affiliate of World Care, an international non-profit 501(c)(3) organization. As of 2006, CRN's Director of Research is Chris Phoenix, and its Executive Director is Mike Treder.

In August 2005, a task force of more than sixty international experts from various fields was organized by CRN to develop comprehensive recommendations for the safe and responsible use of nanotechnology. A series of eleven essays written by members of this CRN Task Force was published in March 2006 by the journal Nanotechnology Perceptions. The Center has sometimes been criticised for concentrating too heavily on the dangers posed by nanotechnology.

National Institute for Nanotechnology

The National Institute for Nanotechnology (NINT) is a Canadian research institution located on the University of Alberta main campus, in Edmonton, Alberta, Canada. Its primary purpose is nanotechnological research.

The institute was established in 2001 as a partnership between the National Research Council of Canada and the University of Alberta, and is funded by the Government of Canada, the Government of Alberta and the university. In June 2006, the institute moved into their present 20,000 square metre facility, designed to be one of the world's largest buildings for nanotechnological research. There are at most two or three other facilities worldwide matching the new building in scale and capacity. According to NINT's research plan, the institute's research focus is in the following areas:

- o Synthesis and characterization of nanocrystals and nanowires
- o Synthesis of supramolecular-based nanomaterials
- o Fabrication and characterization of molecular-scale devices and nanosensors
- o Development of nano-scaled materials for catalysis and directed chemical reactions at semiconductor surfaces
- o Development of nano-electronic and nano-fluidic systems to interface devices to the outside world
- o Theory, modelling, and simulation of nanosystems on multiple length scales
- o Development of quantitative imaging and characterization techniques that support nanotechnology research

The work is focused around four interdisciplinary research groups:

- o Molecular scale development
- o Supramolecular nanoscale assembly
- o Materials and interfacial chemistry
- o Theory and modelling

COMPLEX NANOMATERIALS

To understand complex nanomaterials or nanomaterials having complex composition and their utility, a short narrative on nanomaterials is provided.

Nanomaterials are materials possessing one or more dimensional features having a length on the order of a billionth of a meter to less than 100 billionths of a meter. They are important because (1) they exhibit unique properties which are derived from the size of these features and (2) we have fairly recently learned how to manipulate matter on these dimensions so as to understand and exploit their unique properties and their relationship to size. The properties of matter are largely a result of their composition, and the conditions under which they were produced.

The properties of most materials are largely a result of variations in composition, temperature and pressure and the rates at which these independent parameters are changed during their synthesis or production. As nanotechnologists, we look at "size" as an independent degree of freedom which can be manipulated independent of composition, temperature and pressure to yield materials that possess new properties not exhibited by their conventional counterparts. When materials possesses size features that are on the order of a few billionths of a meter, those materials often exhibit new properties not found in their ordinary material counterparts.

The NNI website which puts these ideas in a simple phrase, "At the nanoscale, the physical, chemical, and biological properties of materials differ in fundamental and valuable ways from the properties of individual atoms and molecules or bulk matter." This phrase can be further simplified to read: At the nanoscale, physical, chemical, and biological properties differ from the properties of individual atoms and molecules or bulk matter.

The nuances of how a material is produced often affects the molecular structure and the possible structure so produced give rise to variations in the materials bulk properties. An example of this is the case hardening of steel. A hard skin is created over a softer core by heating the steel to a high temperature and then rapidly cooling (quenching) it. The resulting composite structure, hard skin/softer core that develops is a function of the rate at which the temperature changed on the surface as compared to that in the core. (This will be discussed in greater detail below.)

To understand nanomaterials, normal materials (non-nanomaterials) are described to provide context. Materials are composed of atoms, molecules, ions, compositions thereof and assemblies of other materials. Solid materials can be broadly classified into crystalline and amorphous.

Crystalline materials come in two main varieties, single crystalline and polycrystalline. A crystalline material is composed of an orderly array of atoms, molecules or ions. Crystalline materials generally have short and long range order. That means that the manner in which atoms are arranged at any one location within a crystal is identical to the arrangement of at any other location. An example of a crystalline material is ordinary table salt. Each grain of salt is a single crystal. A diamond is also a single crystal. The semiconductor or chip inside a computer's microprocessor is a single crystal of silicon. An example of a polycrystalline material is a ceramic dish or coffee cup. If the broken edge of such item is examined with a magnifying glass one can easily see the individual crystals, referred to as crystallites or grains.

When a metal breaks, the grains or crystallites of which it is composed can be observed. In a figure coming soon is a metallograph of a polished metal surface. The grains are easily observed. The surface between grains is called the "grain boundary," and they appear as the line between grains in Figure coming soon. All crystalline materials exhibit Bragg scattering of x-rays. Amorphous materials are non crystalline. They have short range order but lack long range order. Polymers such as polyethylene, polyester and polypropylene are amorphous. Glass is an amorphous solid, which has short range order (always a Si next to an O, etc.) but lacks long range order. Amorphous materials are usually not composed of grains.

Nanomaterials include materials where the size of the particles, crystallites or grains of which the material is composed are on the order of nanometers. As a result of this properties emerge that are not characteristic of their ordinary counterparts having the same composition. These properties (examples given below) some from the fact that within each grain there is a relatively small number of atoms, molecules or ions as compared to a normal material. As the particle or grain size is in the nanometer range, many of the atoms or each grain or within the material as a whole reside at the particle surface, or grain boundary. The number of particle that reside at the surface or grain boundary is proportional to 1/r where r is the average radius of the particle size or grain.

That means that, as the particle size becomes smaller the number of atoms, molecules or ions that exist at the surface or grain boundary increases. Accordingly, the properties of a large collection of nanomaterials are dominated by the properties of the grain boundary or surface. As an example of how the size affects properties consider a material that exhibits ordered electronic spin on the surface but disordered electronic spin in the interior of the particle.

As the particle is reduced in size, the number of atoms on the surface increases as compared to those in the interior and so the number of ordered electronic spins increases as the particle size decreases. A material completely composed of nanoparticles or nano-crystallites of this type would exhibit properties of derived from the ordered spins, e.g. magnetism, while a normal material having the same composition would not exhibit such a property.

Nanomaterials can exist in several forms. The simplest are nanopowders. Nanopowders can be consolidated so as to form a polycrystalline nanomaterials. They can also be combined with other materials to form composite materials.

COMPARATIVE ANALYSIS

Complex materials are analogous to steel or other metal alloys because of their enhanced performance characteristics, while simple materials are more analogous to pure iron, aluminum or copper. An alloy is a mixture of several elements such as in the case of stainless steel: iron, nickel and chromium, each component of which imparts important characteristics to the performance of the alloy. A pure metal is predominantly composed of a single element, for example iron, as in an iron skillet or heating radiator found in homes built in the first half of the 20th century in this country. The first train rails were made out of iron, while all modern rails use steel alloys. Why? Because modern rails last longer, carry many times more weight, are less susceptible to corrosion, and will not become brittle and crack with changing temperatures. Virtually all modern metals in ordinary use are made out of alloys that improve performance, strength and longevity.

Early successes in the nanomaterials industry (use of nano particles in sun screen and ultra fine polishing of semiconductor wafers) represent simple material applications and in these applications only rudimentary aspects of nanotechnology are exploited. Based on current performance limitations, future developments in fuel cells, solar technology, electronics, optics, biotechnology, cosmetics and advanced structural materials involve solutions requiring complex nanomaterials.

Nanotoxicology

Nanotoxicology is the study of the toxicity of nanomaterials. Because of the small size and large surface area of nanomaterials, these materials have unique properties compared with their larger counterparts. The nanomaterials, even when they are made of inert elements like gold, become very active at a nanometer range. Nanotoxicological studies are intended to determine whether and to what extent these may pose a threat to the environment and to human beings.

Research on ultrafine particles has laid the foundation for the emerging field of nanotoxicology, with the goal of studying the biokinetics of engineered nanomaterials and their potential for causing adverse effects. Most reports find that ultrafine particles are more toxic than equivalent larger-sized particles of a given material at similar doses per gram of body weight.

Assessing the safety of engineered nanoparticles is a highly complex matter that goes beyond traditional toxicology. Engineered nanoparticles are not a uniform group of substances. The problem arises from the fact that particle size alone (which determines the surface area of a given mass of a substance) is not the only factor that determines the toxicological impact of a material and that makes nanoparticles of a given substance behave differently from the bulk form.

While the release and production of nanoparticles during industrial and combustion processes and activities is mostly unintentional, the emergence of nanotechnological production processes introduces the intentional and controlled manufacture of nanoparticles. The latter can be further differentiated as either bulk nanoparticles in industry, e.g., carbon black or titanium dioxide, or so-called engineered nanoparticles, e.g., carbon nanotubes.

Given the prospects for nanotechnology, and the fact that products containing engineered nanoparticles have already been introduced to the marketplace, the increasing flow of new products will bring about the massive production of engineered nanoparticles.

Despite all the scientific knowledge gained in the toxicology of particulate matter, because of the many variables involved, scientists still cannot accurately predict how nanomaterials will affect living organisms. What is clear, though, is that the biologic activity and biokinetics of nanoparticles are different from larger particles, and that they depend on many parameters. These parameters can modify cellular uptake, protein binding, translocation from portal of entry to target site, and the possibility of tissue injury.

Important from a toxicological point of view are the physiochemical properties that come with size and lead to certain biological reactivity. Size, together with differences in shape, surface structure, chemical composition (purity, crystallinity, electrical properties, etc.), solubility and biopersistence make for a large number of variables that need to be considered when assessing nanoparticles. On top of that, powders or liquids containing nanoparticles are almost never monodisperse but contain a range of particle sizes. This complicates experimental analysis as larger nanoparticles might have different properties than smaller ones. Also, nanoparticles show a tendency to aggregate and such aggregates often behave differently from individual nanoparticles.

Most, if not all, toxicological studies on nanoparticles rely on current methods, practices and terminology as gained and applied in the analysis of micro-and ultrafine particles and mineral fibers. Together with recent studies on nanoparticles, this provides an initial basis for evaluating the primary issues in a risk assessment framework for nanomaterials.

Given the many parameters involved, nanotoxicology requires an interdisciplinary team approach, even more so than classical toxicology, in order to arrive at an appropriate risk assessment.

As a still-maturing science, nanotoxicology will expand the boundaries of traditional toxicology from a testing and auxiliary science to a new discipline where toxicological knowledge of nanomaterials can be put to constructive use in therapeutics as well as the development of new and better biocompatible materials.

NANOTRIBOLOGY

Nanotribology is a branch of tribology which studies friction phenomenon at the nanometer scale. The distinction between nanotribology and tribology is primarily due to the involvement of atomic forces in the determination of the final behavior of the system.

Gears, bearings, and liquid lubricants can reduce friction in the macroscopic world, but the origins of friction for small devices such as micro-or nanoelectromechanical systems (NEMS) require other solutions. Despite the unprecedented accuracy by which these devices are nowadays designed and fabricated, their enormous surface-volume ratio leads to severe friction and wear issues, which dramatically reduce their applicability and lifetime. Traditional liquid lubricants become too viscous when confined in layers of molecular thickness. This situation has led to a number of proposals for ways to reduce friction on the nanoscale, such as superlubricity and thermolubricity.

NANOPHOTONICS

Nanophotonics is the study of the behavior of light on the nanometre scale. The ability to fabricate devices in nanoscale that has been developed recently provided the catalyst for this area of study.

Nanophotonics scientists are experimenting with different ways to generate and manipulate light using ultrasmall, engineered structures at the nanoscale. Tiny gold particles called "nanostars" are being studied to ascertain how they interact with light. The study has the potential to revolutionise the telecommunications industry by providing low power, high speed, interference-free devices such as electrooptic and all-optical switches on a chip. There are a number of research groups working in this area in the U.S., the UK, Italy and Spain.

MECHANOCHEMISTRY

Mechanochemistry, is the coupling of the mechanical and the chemical on a molecular scale includes mechanical breakage, polymer degradation under shear, cavitation related phenomena (e.g., sonochemistry and sonoluminescence), shockwave chemistry and physics, and even the burgeoning field of molecular machines.

A small part of mechanochemistry is sometimes also called "positional synthesis" or "positional assembly" is a technique for forming chemical bonds by direct computer control of the position of molecules. This is an example of a specific type of Mechanosynthesis.

As of 2004, the typical experimental arrangement is to attach a molecule to the tip of an atomic force microscope, and then use the microscope's precise positioning abilities to push the molecule on the tip into another on a substrate. Since the angles and distances can be precisely controlled, and the reaction occurs in a vacuum, novel chemical compounds and arrangements are possible. Much of the excitement regarding mechanochemistry regards its potential use in automated assembly of molecular-scale devices. Such techniques appear to have many applications in medicine, aviation, resource extraction, manufacturing and warfare.

Most theoretical explorations of such machines have focused on using Carbon, because of the many strong bonds it can form, the many types of chemistry these bonds permit, and utility of these bonds in medical and mechanical applications. Carbon forms diamond, for example, which if cheaply available, would be an excellent material for many machines. In practice, getting exactly one molecule to a known place on the microscope's tip is possible, but has proven difficult to automate. Since practical products require at least several hundred million atoms, this technique has not yet proven practical in forming a real product.

The goal of mechanoassembly research at this point focuses on overcoming these problems by calibration, and selection of appropriate synthesis reactions. The first product to be built by these means will probably be a specialized, very small (roughly 1,000 nanometers on a side) machine tool that can build copies of itself using mechanochemical means, under the control of an external computer. In the literature, such a tool is called an assembler. Once assemblers exist, geometric growth (copies making copies) could reduce the cost of assemblers rapidly. Control by an external computer should then permit large groups of assemblers to construct large, useful projects to atomic precisions. One such project would combine molecular-level conveyor belts with permanently-mounted assemblers to produce a factory.

Mechanochemistry actually goes back more than 100 years. The coupling of the mechanical and the chemical on a molecular scale includes mechanical breakage, polymer degradation under shear, cavitation related phenomena (e.g., sonochemistry and sonoluminescence), shockwave chemistry and physics, and even the burgeoning field of molecular machines.

The technique of moving single atoms mechanically was proposed by Eric Drexler in his 1986 book The Engines of Creation. In 1988, researchers at IBM's Zürich Research Institute successfully spelled the letters "IBM" in Xenon atoms on a cryogenic copper surface, grossly validating the approach. Since then, a number of research projects have undertaken to use similar techniques to store computer data in a compact fashion. More recently the technique has been used to explore novel physical chemistries, sometimes using lasers to excite the tips to particular energy states, or examine the quantum chemistry of particular chemical bonds.

NANO-OPTICS

Nano-optics is the branch of optical engineering which deals with optics at deeply subwavelength length scales. Technologies in the realm of nano-optics include near-field scanning optical microscopy (NSOM), photoassisted scanning tunnelling microscopy, and surface plasmon optics. Traditional microscopy makes use of diffractive elements to focus light tightly in order to increase resolution.

But because of the diffraction limit (also known as the Rayleigh Criterion), propagating light may be focused to a spot with a minimum diameter of roughly half the wavelength of light. Thus, even with diffraction-limited confocal microscopy, the maximum resolution obtainable is on the order of a couple hundred nanometers. The scientific and industrial communities are becoming more interested in the characterization of materials and phenomena on the scale of a few nanometers, so alternative techniques must be utilized. Scanning Probe Microscopy (SPM) makes use of a "probe," (usually either a tiny aperture or super-sharp tip), which either locally excites a sample or transmits local information from a sample to be collected and analyzed.

NANOLANGUAGE

NanoLanguage is a new way of thinking scientific computing, combining the strength of flexible object-oriented scripting interfaces (known from Mathematica and MatLab) with sophisticated high performance scientific computing algorithms. The goal is to enable scientists to efficiently extend, specialize and combine methods to calculate nanoscale properties of matter, including density functional theory, semi-empirical tight-binding, classical potentials, k.p and various quantum-chemical methods.

NanoLanguage allows for both low and high level detailed control of the computer simulations performed in the Atomistix ToolKit. At the high level, it offers a common interface for setting up complex atomic-scale simulations and analyzing the results. On the lower level side, it provides an interface to the low-level functionality in ATK.

NanoLanguage is built on top of Python, a powerful and well-established interpreted programming language, and thus it includes basic elements such as loops over simulation control parameters, plus support for efficient manipulations of e.g. numerical array data. It is therefore an ideal tool for automating series of simulations where geometric, material, or other parameters are to be optimized.

Using NanoLanguage allows scientists to express models of nature in a common language without the need to re-implement already implemented algorithms, and it allows for third-party development of new functionality on top of the ATK platform. Such functionality may consist of new atomic-scale modeling methodologies, tailored semi-empirical methods, or complex post-processing methods for calculating new quantities from the fundamental simulation results.

NANOCRYSTAL SOLAR CELL

Nanocrystal solar cells or quantum dot solar cells, are solar cells based on a silicon substrate with a coating of nanocrystals. Whilst previous methods of quantum dot creation relied on expensive molecular beam epitaxy processes, fabrication using colloidal synthesis allows for a more cost effective manufacture. A thin film of nanocrystals is obtained by a process known as "spin-coating". This involves placing an amount of the quantum dot solution onto a flat substrate, which is then rotated very quickly. The solution spreads out uniformly, and the substrate is spun until the required thickness is achieved.

Quantum dot based photovoltaic cells based around dye-sensitised colloidal TiO2 films were investigated in 1991 and were found to exhibit promising efficiency of converting incident light energy to electrical energy, and were found to be incredibly encouraging due to the low cost of materials in the search for more commercially viable/affordable renewable energy sources.

Although research is still in its infancy and is ongoing, in the future quantum dot based photovoltaics may offer advantages such as mechanical flexiblity (quantum dot-polymer composite photovoltaics) as well as low cost, clean power generation.

14

THE NANOMANUFACTURING AND ITS PROCEDURE

Nanomanufacturing is used to describe either the production of nano scaled materials, which can be powders or fluids, or describe the manufacturing of parts "button up" from nanoscaled materials or "top down" in smallest steps for high precison by several technologis like laser ablation, etching and others. The term is widely used, e.g. by the European Technology Platform MINAM and the U.S. National Nanotechnology Initiative (NNI). NNI refer to a sub-effort of nanotechnology--one of its five "priority areas."

There is also a Nanomanufacturing Program at the U.S. National Science Foundation. Nanomanufacturing should not be confused with molecular manufacturing which refers specifically to the manufacture of complex, nanoscale structures by means of nonbiological mechanosynthesis (and subsequent assembly). The NNI has defined nanotechnology very broadly, to include a wide range of tiny structures including those created by large and imprecise tools.

Nanomanufacturing is not defined in the NNI's recent report, "Instrumentation and Metrology for Nanotechnology." (For contrast, another "priority area," nanofabrication, is defined as "the ability to fabricate, by directed or self-assembly methods, functional structures or devices at the atomic or molecular level." p. 67) Nanomanufacturing appears to be the near-term industrial-scale manufacture of nanotechnology-based objects, with emphasis on low cost and reliability.

CATALYST BASED ON NANOMATERIAL

Nanomaterial-based catalysts are usually heterogeneous catalysts broken up into nanoparticles in order to speed up the catalytic process. The extremely small size of the particles maximizes the surface area exposed to the reactant, allowing more reactions to occur at the same time, thus speeding up the process. Much research on nanomaterial-based catalysts have to do with maximizing the effectiveness of the catalyst coating in fuel cells. Platinum is currently the most common catalyst for this application, however, it is expensive and rare, so a lot of research has been going into maximizing the catalytic properties of other metals by shrinking them to nanoparticles in the hope that someday they will be an efficient and economic alternative to platinum.

Defense programs in many countries are now concentrating on nanotechnology research that will facilitate advances in such technology used to create secure but small messaging equipment, allow thc development of smart weapons, improve stealth capabilities, aid in developing specialized sensors (including bio-inclusive sensors), help to create self-repairing military equipment, and improve the development and delivery mechanisms for medicines and vaccines. Nanotechnology builds on advances in microelectronics during the last decades of the twentieth century. The miniaturization of electrical components greatly increased the utility and portability of computers, imaging equipment, microphones, and other electronics.

Indeed, the production and wide use of such commonplace devices such as personal computers and cell phones was absolutely dependent on advances in microtechnology.

Despite these fundamental advances there remain real physical constraints (e.g., microchip design limitations) to further miniaturization based upon conventional engineering principles. Nanotechnologies intend to revolutionize components and manufacturing techniques to overcome these fundamental limitations. In addition, there are classes of biosensors and feedback control devices that require nanotechnology because-despite advances in microtechnology-present components remain too large or slow.

ADVANCES

Nanotechnology advances affect all branches of engineering and science that deal directly with device components ranging in size between 1/10,000,000 (one ten millionth of a millimeter) and 1/10,0000 millimeter. At these scales, even the most sophisticated microtechnology-based instrumentation is useless. Engineers anticipate that advances in nanotechnology will allow the direct manipulation of molecules in biological samples (e.g., proteins or nucleic acids) paving the way for the development of new materials that have a biological component or that can provide a biological interface.

In addition to new tools, nanotechnology programs advance practical understanding of quantum physics. The internalization of quantum concepts is a necessary component of nanotechnology research programs because the laws of classical physics (e.g., classical mechanics or generalized gas laws) do not always apply to the atomic and near-atomic level. Nanotechnology and quantum physics. Quantum theory and mechanics describe the relationship between energy and matter on the atomic and subatomic scale. At the beginning of the twentieth century, German physicist Maxwell Planck (1858-1947) proposed that atoms absorb or emit electromagnetic radiation in bundles of energy termed quanta. This quantum concept seemed counter-intuitive to well-established Newtonian physics. Advancements associated with quantum mechanics (e.g., the uncertainty principle) also had profound implications with regard to the philosophical scientific arguments regarding the limitations of human knowledge.

Planck's quantum theory, which also asserted that the energy of light (a photon) was directly proportional to its frequency, proved a powerful concept that accounted for a wide range of physical phenomena. Planck's constant relates the energy of a photon with the frequency of light. Along with the constant for the speed of light, Planck's constant (h = 6.626 x 10-34 Joule-second) is a fundamental constant of nature.

Prior to Planck's work, electromagnetic radiation (light) was thought to travel in waves with an infinite number of available frequencies and wavelengths. Planck's work focused on attempting to explain the limited spectrum of light emitted by hot objects. Danish physicist Niels Bohr (1885-1962) studied Planck's quantum theory of radiation and worked in England with physicists J. J. Thomson (1856-1940), and Ernest Rutherford (1871-1937) to improve their classical models of the atom by incorporating quantum theory. During this time, Bohr developed his model of atomic structure.

According to the Bohr model, when an electron is excited by energy it jumps from its ground state to an excited state (i.e., a higher energy orbital). The excited atom can then emit energy only in certain (quantized) amounts as its electrons jump back to lower energy orbits located closer to the nucleus. This excess energy is emitted in quanta of electromagnetic radiation (photons of light) that have exactly the same energy as the difference in energy between the orbits jumped by the electron.

The electron quantum leaps between orbits proposed by the Bohr model accounted for Plank's observations that atoms emit or absorb electromagnetic radiation in quanta. Bohr's model also explained many important properties of the photoelectric effect described by Albert Einstein (1879-1955). Einstein assumed that light was transmitted as a stream of particles termed photons. By extending the well-known wave properties of light to include a treatment of light as a stream of photons, Einstein was able to explain the photoelectric effect. Photoelectric properties are key to regulation of many microtechnology and proposed nanotechnology level systems.

Quantum mechanics ultimately replaced electron "orbitals" of earlier atomic models with allowable values for angular momentum (angular velocity multiplied by mass) and depicted electron positions in terms of probability "clouds" and regions.

In the 1920s, the concept of quantization and its application to physical phenomena was further advanced by more mathematically complex models based on the work of the French physicist Louis Victor de Broglie (1892-1987) and Austrian physicist Erwin Schrödinger (1887-1961) that depicted the particle and wave nature of electrons. De Broglie showed that the electron was not merely a particle but a waveform. This proposal led Schrödinger to publish his wave equation in 1926. Schrödinger's work described electrons as a "standing wave" surrounding the nucleus, and his system of quantum mechanics is called wave mechanics. German physicist

Max Born (1882-1970) and English physicist P. A. M. Dirac (1902-1984) made further advances in defining the subatomic particles (principally the electron) as a wave rather than as a particle and in reconciling portions of quantum theory with relativity theory.

Working at about the same time, German physicist Werner Heisenberg (1901-1976) formulated the first complete and self-consistent theory of quantum mechanics. Matrix mathematics was well established by the 1920s, and Heisenberg applied this powerful tool to quantum mechanics. In 1926, Heisenberg put forward his uncertainty principle which states that two complementary properties of a system, such as position and momentum, can never both be known exactly. This proposition helped cement the dual nature of particles (e.g., light can be described as having both wave and particle characteristics). Electromagnetic radiation (one region of the spectrum that comprises visible light) is now understood to have both particle and wave like properties.

In 1925, Austrian-born physicist Wolfgang Pauli (1900-1958) published the Pauli exclusion principle states that no two electrons in an atom can simultaneously occupy the same quantum state (i.e., energy state). Pauli's specification of spin (+1/2 or-1/2) on an electron gave the two electrons in any suborbital differing quantum numbers (a system used to describe the quantum state) and made completely understandable the structure of the periodic table in terms of electron configurations (i.e., the energy-related arrangement of electrons in energy shells and suborbitals). In 1931, American chemist Linus Pauling published a paper that used quantum mechanics to explain how two electrons, from two different atoms, are shared to make a covalent bond between the two atoms. Pauling's work provided the connection needed in order to fully apply the new quantum theory to chemical reactions. Advances in nanotechnology depend upon an understanding and application of these fundamental quantum principles. At the quantum level the smoothness of classical physics disappears and nanotechnologies are predicated on exploiting this quantum roughness.

THE APPLICATIONS

The development of devices that are small, light, self-contained, use little energy and that will replace larger microelectronic equipment is one of the first goals of the anticipated nanotechnology revolution. The second phase will be marked by the introduction of materials not feasible at larger than nanotechnology levels. Given the nature of quantum variance, scientists theorize that single molecule sensors can be developed and that sophisticated memory storage and neural-like networks can be achieved with a very small number of molecules. Traditional engineering concepts undergo radical transformation at the atomic level. For example, nano-technology motors may drive gears, the cogs of which are composed of the atoms attached to a carbon ring. Nanomotors may themselves be driven by oscillating magnetic fields or high precision oscillating lasers.

Perhaps the greatest promise for nanotechnology lies in potential biotechnology advances. Potential nano-level manipulation of DNA offers the opportunity to radically expand the horizons of genomic medicine and immunology. Tissue-based biosensors may unobtrusively be able to monitor and regulate site-specific medicine delivery or regulate physiological processes. Nanosystems might serve as highly sensitive detectors of toxic substances or used by inspectors to detect traces of biological or chemical weapons. In electronics and computer science, scientists assert that nanotechnologies will be the next major advance in computing and information processing science. Microelectronic devices rely on recognition and flips in electron gating (e.g. where differential states are ultimately represented by a series of binary numbers ["0" or "1"] that depict voltage states). In contrast, future quantum processing will utilize the identity of quantum states as set forth by quantum numbers. In quantum cryptography systems with the ability to decipher encrypted information will rely on precise knowledge of manipulations used to achieve various atomic states.

Nanoscale devices are constructed using a combination of fabrication steps. In the initial growth stage, layers of semiconductor materials are grown on a dimension limiting substrate. Layer composition can be altered to control electrical and/or optical characteristics. Techniques such as molecular beam epitaxy (MBE) and metallo-organic chemical vapor deposition (MOCVD) are capable of producing layers of a few atoms thickness. The developed pattern is then imposed on successive layers (the pattern transfer stage) to develop desired three dimensional structural characteristics.

NANOTECHNOLOGY RESEARCH

In the United States, expenditures on nanotechnology development tops $500 million per year and is largely coordinated by the National Science Foundation and Department of Defense Advanced Research Projects Agency (DARPA) under the umbrella of the National Nano-technology Initiative. Other institutions with dedicated funding for nanotechnology include the Department of Energy (DOE) and National Institutes of Health (NIH).

Research interests. Current research interests in nano-technology include programs to develop and exploit nanotubes for their ability to provide extremely strong bonds. Nanotubes can be flexed and woven into fibers for use in ultrastrong-but also ultralight-bulletproof vests. Nanotubes are also excellent conductors that can be used to develop precise electronic circuitry.

Other interests include the development of nanotechnology-based sensors that allow smarter autonomous weapons capable of a greater range of adaptations enroute to a target; materials that offer stealth characteristics across a broader span of the electromagnetic spectrum; self-repairing structures; and nanotechnology-based weapons to disrupt-but not destroy-electrical system infrastructure.

PROGRESS OF NANOTECHNOLOGY

A basic question about nanotechnology is, "When will it be achieved?" The answer is simple: No one knows. How molecular machines will behave is a matter for calculation, but how long it will take us to develop them is a separate issue. Technology timetables can't be calculated from the laws of nature, they can only be guessed at. In this chapter, we examine different paths to nanotechnology, hear what some of the pioneers have to say, and describe the progress already made. This will not answer our basic question, but it will educate our guesses.

Molecular nanotechnology could be developed in any of several basically different ways. Each of these basic alternatives itself includes further alternatives. Researchers will be asking, "How can we make the fastest progress?" To understand the answers they may come to, we need to ask the same question here, adopting (for the moment) a gung-ho, let's-go, how-do-we-get-the-job-done? attitude. We give some of the researchers' answers in their own words.

Like "When will it be achieved?", this is a basic question with an answer beyond calculation. Here, though, the answer seems fairly clear. Throughout history, people have worked to achieve better control of matter, to convince atoms to do what we want them to do. This has gone on since before people learned that atoms exist, and has accelerated ever since. Although different industries use different materials and different tools and methods, the basic aim is always the same. They seek to make better things, and make them more consistently, and that means better control of the structure of matter. From this perspective, nanotechnology is just the next, natural step in a progression that has been under way for millennia.

Consider the compact discs now replacing older stereo records: both the old and the new technologies stamp patterns into plastic, but for CDs, the bumps on the stamping surface are only about 130 by 600 nanometers in size, versus 100,000 nanometers or so for the width of the groove on an old-style record. Or look at a personal computer.

John Foster, a physicist at IBM's Almaden Research Center, points to a hard disk and says that "inside that box are a bunch of whirring disks, and every one of those disks has got a metal layer where the information is stored. The last thing on top of the metal layer is a monolayer that's the lubricant between the disk and the head that flies over it. The monolayer is not fifteen angstroms [15 angstroms = 1.5 nanometers] and it's not three, because fifteen won't work and neither will three. So it has to be ten plus or minus a few angstroms. This is definitely working in the nanometer regime. We're at that level: We ship it every day and make money on it every day."

The transistors on computer chips are heading down in size on an exponential curve. Foster's colleague at

IBM, Patrick Arnett, expects the trend to continue: "If you stay on that curve, then you end up at the atomic scale at 2020 or so. That's the nature of technology now. You expect to follow that curve as far as you can go."

The trend is clear, and at least some of the results can be foreseen, but the precise path and timetable for the development of nanotechnology is unpredictable. This unpredictability goes to the heart of important questions: "How will this technology be developed? Who will do it? Where? When? In ten years? Fifty? A hundred? Will this happen in my lifetime?" The answers will depend on what people do with their time and resources, which in turn will depend on what goals they think are most promising. Human attitudes, understanding, and goals will make all the difference.

DECISIONS AFFECTING THE RATE OF ADVANCE

Decisions about research directions are central. Researchers are already pouring effort into chemical synthesis, molecular engineering, and related fields. The same amount of effort could produce more impressive results in molecular nanotechnology if a fraction of it were differently directed. The research funders-corporate executives, and decision makers in science funding agencies like the National Science Foundation in the United States and Japan's Ministry of International Trade and Industry-all have a large influence on research directions, but so do the researchers working in the labs. They submit proposals to potential funders (and often spend time on personally chosen projects, regardless of funding), so their opinions also shape what happens. Where public money is involved, politicians' impressions of public opinion can have a huge influence, and public opinion depends on what all of us think and say..

Still, researchers play a central role. They tend to work on what they think is interesting, which depends on what they think is possible which depends on the tools they have or-among the most creative researchers-on the tools they can see how to make. Our tools shape how we think: as the saying goes, when all you have is a hammer, everything looks like a nail. New tools encourage new thoughts and enable new achievements, and decisions about tool development will pace advances in nanotechnology. To understand the challenges ahead, we need to take a look at ideas about the tools that will be needed.

IMPORATANCE OF TOOLS

Throughout history, limited tools have limited achievement. Leonardo da Vinci's sixteenth century chain drives and ball bearings were theoretically workable, yet never worked in their inventor's lifetime. Charles Babbage's nineteenth century mechanical computer suffered the same fate. The problem? Both inventors needed precisely machined parts that (though readily available today) were beyond the manufacturing technology of their times. Physicist David Miller recounts how a sophisticated integrated circuit design project at TRW hit similar limits in the early 1980s: "It all came down to whether a German company could cool their glass lenses slowly enough to give us the accuracy we needed. They couldn't."

In the molecular world, tool development again paces progress, and new tools can bring breathtaking advances. Mark Pearson, director of molecular biology for Du Pont, has observed this in action: "When I was a graduate student back in the 1950s, it was a multiyear problem to determine the molecular structure of a single protein. We used to say, 'one protein, one career.' Yet now the time has shrunk from a career to a decade to a year-and in optimal cases to a few months." Protein structures can be mapped atom by atom by studying X-ray reflections from layers in protein crystals. Pearson observes that "Characterizing a protein was a career-long endeavor in part because it was so difficult to get crystals, and just getting the material was a big constraint. With new technologies, we can get our hands on the material now-that may sound mundane, but it's a great advance. To the people in the field, it makes all the difference in the world." Improved tools for making and studying proteins are of special importance because proteins are promising building blocks for first-generation molecular machines.

Science Discovery

Nobel Prizes are more often awarded for discoveries than for the tools (including instruments and techniques) that made them possible. If the goal is to spur scientific progress, this is a shame. This pattern of reward extends throughout science, leading to a chronic underinvestment in developing new tools. Philip Abelson, an editor of the journal Science, points out that the United States suffers from "a lack of support for development of new instrumentation. At one time, we had a virtual monopoly in pioneering advances in instrumentation. Now practically no federal funds are available to universities for the purpose." It's easier and less risky to squeeze one more piece of data out of an existing tool than to pioneer the development of a new one, and it takes less imagination.

But new tools emerge anyway, often from sources in other fields. The study of protein crystals, for example, can benefit from new X-ray sources developed by physicists, and techniques from chemistry can help make new proteins. Because they can't anticipate tools resulting from innovations in other fields, scientists and engineers are often too pessimistic about what can be achieved in their own fields. Nanotechnology will join several fields, and yield tools useful in many others. We should expect surprising results.

RESEARCHERS' TOOLS

Today's tools for making small-scale structures are of two kinds: molecular-processing tools and bulk-processing tools. For decades, chemists and molecular biologists have been using better and better molecular-processing tools to make and manipulate precise, molecular structures. These tools are of obvious use. Physicists, as we will see, have recently developed tools that can also manipulate molecules. Combined with techniques from chemistry and molecular biology, these physicist's tools promise great advances. Microtechnologists have applied chip-making techniques to the manufacture of microscopic machines. These technologies-the main approach to miniaturization in recent decades-can play at most a supporting role in the development of nanotechnology. Despite appearances, it seems that microtechnology cannot be refined into nanotechnology.

SMALL MICROTECHNOLOGY

For many years, it was conventional to assume that the road to very small devices led through smaller and smaller devices: a top-down path. On this path, progress is measured by miniaturization: How small a transistor can we build? How small a motor? How thin a line can we draw on the surface of a crystal? Miniaturization focuses on scale and has paid off well, spawning industries ranging from watchmaking to microelectronics.

Researchers at AT&T Bell Labs, the University of California at Berkeley, and other laboratories in the

United States have used micromachining (based on microelectronic technologies) to make tiny gears and even electric motors. Micromachining is also being pursued successfully in Japan and Germany. These microgears and micromotors are, however, enormous by nanotechnological standards: a typical device is measured in tens of micrometers, billions of times the volume of comparable nanogears and nanomotors. (In our simulated molecular world, ten microns is the size of a small town.) In size, confusing microtechnology with molecular nanotechnology is like confusing an elephant with a ladybug.

The differences run deeper, though. Microtechnology dumps atoms on surfaces and digs them away again in bulk, with no regard for which atom goes where. Its methods are inherently crude. Molecular nanotechnology, in contrast, positions each atom with care. As Bill DeGrado, a protein chemist at Du Pont, says, "The essence of nanotechnology is that people have worked for years making things smaller and smaller until we're approaching molecular dimensions. At that point, one can't make smaller things except by starting with molecules and building them up into assemblies." The difference is basic: In microtechnology, the challenge is to build smaller; in nanotechnology, the challenge is to build bigger-we can already make small molecules.

A language warning: in recent years, nanotechnology has indeed been used to mean "very small microtechnology"; for this usage, the answer to the above question is yes, by definition. This use of a new word for a mere extension of an old technology will produce considerable confusion, particularly in light of the widespread use of nanotechnology in the sense found here. Nanolithography, nanoelectronics, nanocomposites, nanofabrication: not all that is nano- is molecular, or very relevant to the concerns raised in this book. The terms molecular nanotechnology and molecular manufacturing are more awkward but avoid this confusion.

MICROTECHNOLOGY LEADING TO NANOTECHNOLOGY

Can bulldozers can be used to make wristwatches? At most, they can help to build factories in which watches are made. Though there could be surprises, the relevance of microtechnology to molecular nanotechnology seems similar. Instead, a bottom-up approach is needed to accomplish engineering goals on the molecular scale.

Tools Used for Molecular Engineering

Almost by definition, the path to molecular nanotechnology must lead through molecular engineering. Working in different disciplines, driven by different goals, researchers are making progress in this field. Chemists are developing techniques able to build precise molecular structures of sorts never before seen. Biochemists are learning to build structures of familiar kinds, such as proteins, to make new molecular objects.

In a visible sense, most of the tools used by chemists and biochemists are rather unimpressive. They work on countertops cluttered with dishes, bottles, tubes, and the like, mixing, stirring, heating, and pouring liquids-in biochemistry, the liquid is usually water with a trace of material dissolved in it. Periodically, a bit of liquid is put into a larger machine and a strip of paper comes out with a graph printed on it. As one might guess from this description, research in the molecular sciences is usually much less expensive than research in high-energy physics (with its multibillion-dollar particle accelerators) or research in space (with its multibillion-dollar spacecraft). Chemistry has been called "small science," and not because of the size of the molecules. Of these fields, it is biomolecular science that is most obviously developing tools that can build nanotechnology, because biomolecules already form molecular machines, including devices resembling crude assemblers. This path is easiest to picture, and can surely work, yet there is no guarantee that it will be fastest: research groups following another path may well win. Each of these paths is being pursued worldwide, and on each, progress is accelerating.

Physicists have recently contributed new tools of great promise for molecular engineering. These are the proximal probes, including the scanning tunneling microscope (STM) and the atomic force microscope (AFM). A proximal-probe device places a sharp tip in proximity to a surface and uses it to probe (and sometimes modify) the surface and any molecules that may be stuck to it.

MOLECULAR BUILDING BLOCKS

A huge technology base for molecular construction already exists. Tools originally developed by biochemists and biotechnologists to deal with molecular machines found in nature can be redirected to make new molecular machines. The expertise built up by chemists in more than a century of steady progress will be crucial in molecular design and construction. Both disciplines routinely handle molecules by the billions and get them to form patterns by self-assembly. Biochemists, in particular, can begin by copying designs from nature.

Molecular building-block strategies could work together with proximal probe strategies, or could replace them, jumping directly to the construction of large numbers of molecular machines. Either way, protein molecules are likely to play a central role, as they do in nature.

Protein Engineering Building Molecular Machines

Proteins can self assemble into working molecular machines, objects that do something, such as cutting and splicing other molecules or making muscles contract. They also join with other molecules to form huge assemblies like the ribosome (about the size of a washing machine, in our simulation view). Ribosomes-programmable machines for manufacturing proteins-are nature's closest approach to a molecular assembler. The genetic-engineering industry is chiefly in the business of reprogramming natural nanomachines, the ribosomes, to make new proteins or to make familiar proteins more cheaply. Designing new proteins is termed protein engineering. Since biomolecules already form such complex devices, it's easy to see that advanced protein engineering could be used to build first-generation nanomachines.

FANCY MOLECULAR MACHINES

Making proteins is easier than designing them. Protein chemists began by studying proteins found in nature, but have only recently moved on to the problem of engineering new ones. These are called de novo proteins, meaning completely new, made from scratch. Designing proteins is difficult because of the way they are constructed. As Bill DeGrado, a protein chemist at Du Pont, explains: "A characteristic of proteins is that their activities depend on their three-dimensional structures. These activities may range from hormonal action to a function in digestion or in metabolism.

Whatever their function, it's always essential to have a definite three-dimensional shape or structure." This three-dimensional structure forms when a chain folds to form a compact molecular object. To get a feel for how tough it is to predict the natural folding of a protein chain, picture a straight piece of cord with hundreds of magnets and sticky knots along its length. In this state, it's easy to make and easy to understand. Now pick it up, put it in a glass jar, and shake it for a long time. Could you predict its final shape? Certainly not: it's a tangled mess. One might call this effort at prediction "the sticky-cord-folding problem"; protein chemists call theirs "the protein-folding problem."

Given the correct conditions, a protein chain always folds into one special shape, but that shape is hard to predict from just the straightened structure. Protein designers, though, face the different job of first determining a desired final shape, and then figuring out what linear sequence of amino acids to use to make that shape. Without solving the classic protein-folding problem, they have begun to solve the protein-design problem.

THE ACCOMPLISHMENTS SO FAR

Bill DeGrado and his colleagues at Du Pont had one of the first successes: "We've been able to use basic principles to design and build a simple molecule that folds up the way we want it to. This is really the first real example of a designed protein structure, designed from scratch, not by taking an already existing structure and tinkering with it."

Although scientists do the work, the work itself is really a form of engineering, as shown by the title of the field's journal, Protein Engineering. Bill DeGrado's description of the process makes this clear: "After you've made it, the next step is to find out whether your protein did what you expected it to do. Did it fold? Did it pass ions across bilayers [such as cell membranes]? Does it have a catalytic function [speeding specific chemical reactions]? And that's tested using the appropriate experiment. More than likely, it won't have done what you wanted it to do, so you have to find out why. Now, a good design has in it a contingency plan for failure and helps you learn from mistakes. Rather than designing a structure that would take a year or more to analyze, you design it so that it can be assayed for given function or structure in a matter of days."

Many groups are pursuing protein design today, including academic researchers like Jane and Dave Richardson at Duke University, Bruce Erickson at the University of North Carolina, and Tom Blundell, Robin Leatherbarrow, and Alan Fersht in Britain. The successes have started to roll in. Japan, however, is unique in having an organization devoted exclusively to such projects: the Protein Engineering Research Institute (PERI) in Osaka. In 1990, PERI announced the successful design and construction of a de novo protein several times larger than any built before.

Speciality of Proteins

The main advantage of proteins is that they are familiar: a lot is known about them, and many tools exist for working with them. Yet proteins have disadvantages as well. Just because this design work is starting with proteins-soft, squishy molecules that are only marginally suitable for nanotechnology-doesn't mean it will stay within those limits.

De Grado points out "The fundamental goal of our work in de novo design is to be able to take the next step and get entirely away from protein systems." An early example is the work of Wallace Carothers of Du Pont, who used a de novo approach to studying the nature of proteins: Rather than trying to cut up proteins, he tried to build up things starting with amino acids and other similar monomers. In 1935, he succeeded in making nylon.

DeGrado explains "There is a deep philosophical belief at Du Pont in the ability of people to make molecules de novo that will do useful things. And there is a fair degree of commitment from the management that following that path will lead to products: not directly, and not always predictably, but they know that they need to support the basic science.

"I think ultimately we have a better chance at doing some really exciting things by de novo design, because our repertory should be much greater than that of nature. Think about the ability to fly: One could breed better carrier pigeons or one could design airplanes." The biology community, however, leans more toward ornithology than toward aerospace engineering. DeGrado's experience is that "a lot of biologists feel that if you aren't working with the real thing [natural proteins], you aren't studying biology, so they don't totally accept what we're doing. On the other hand, they recognize it as good chemistry."

Where is Protein Engineering Headed?

Like the IBM physicists, protein designers are moved by a vision of molecular engineering. In 1989, Bill DeGrado predicted, "I think we'll be able to make catalysts or enzymelike molecules, possibly ones that catalyze reactions not catalyzed in nature." Catalysts are molecular machines that speed up chemical reactions: they form a shape for the two reacting molecules to fit into and thereby help the reaction move faster, up to a million reactions per second. New ones, for reactions that now go slowly, will give enormous cost savings to the chemical industry.

This prediction was borne out just a few months later, when Denver researchers John Stewart, Karl Hahn, and Wieslaw Klis announced their new enzyme, designed from scratch over a period of two years and built successfully on the first try. It's a catalyst, making some reactions go about 100,000 times faster. Nobel Prize-winning biochemist Bruce Merrifield believes that "if others can reproduce and expand on this work, it will be one of the most important achievements in biology or chemistry."

DeGrado also has longer term plans for protein design, beyond making new catalysts: "It will allow us to think about designing molecular devices in the next five to ten years. It should be possible ultimately to specify a particular design and build it. Then you'll have, say, proteinlike molecules that self-assemble into complex molecular objects, which can serve as machinery. But there's a limit to how small you can make devices. You'll shrink things down so far and then you won't be able to go any further, because you've reached molecular dimensions."

Mark Pearson shows that management at Du Pont also has this vision. Regarding the prospects for nanotechnology and assemblers, he remarked, "You know, it'll take money and effort and good ideas for sure. But to my way of thinking, there is no absolute fundamental limitation to preclude us from doing this kind of thing." He didn't say his company plans to develop nanotechnology, but such plans aren't really necessary. Du Pont is already on the nanotechnology path, for other-shorter-term, commercial-reasons. Like IBM, if they do decide to move quickly, they have the resources and forward-looking people needed to succeed.

Who Else Builds Molecular Objects?

Chemists, most of whom do not work on proteins, are the traditional experts in building molecular objects. As a group they've been building molecules for over a century, with ever increasing ability and confidence. Their methods are all indirect: They work with billions of atoms at a time-massive parallelism-but without control of the positions of their workpieces. The molecules typically tumble randomly in a liquid or gas, like pieces of a puzzle that may or may not fit together correctly when shaken together in a box. With clever design and planning, most pieces will join properly.

Chemists mix molecules on a huge scale (in our simulation view, a test tube holds a churning molecular swarm with the volume of an inland sea), yet they still achieve precise molecular transformations. Given that they work so indirectly, their achievements are astounding. This is, in part, the result of the enormous amount of work poured into the field for many decades. Thousands of chemists are working on molecular construction in the United States alone; add to that the chemists in Europe, in Japan, and in the rest of the world, and you have a huge community of researchers making great strides. Though it publishes only a one-paragraph summary of each research report, a guide to the chemical literature-Chemical Abstracts-covers several library walls and grows by many feet of shelf space every year.

How can Mixing Chemicals Build Molecular Objects?

An engineer would say that chemists (at least those specializing in synthesis) are doing construction work, and would be amazed that they can accomplish anything without being able to grab parts and put them in place. Chemists, in effect, work with their hands tied behind their backs. Molecular manufacturing can be termed "positional chemistry" or "positional synthesis," and will give chemists the ability to put molecules where they want them in three-dimensional space. Rather than trying to design puzzle pieces that will stick together properly by themselves when shaken together in a box, chemists will then be able to treat molecules more like bricks to be stacked. The basic principles of chemistry will be the same, but strategies for construction will become far simpler.

Without positional control, chemists face a problem something like this: Picture a giant glass barrel full of tiny battery-powered drills, buzzing away in all directions, vibrating around in the barrel. Your goal is to take a piece of wood and put a hole in just one specific spot. If you simply throw it in the barrel, it will be drilled haphazardly in many places.

To control the process, you must protect all the places you don't want drilled-perhaps by gluing protective pieces of metal over most of the wood surface. This problem-how to protect one part of a molecule while altering another part-has forced chemists to develop ever-cleverer ploys to build larger and larger molecules.

If Chemists Can Make Molecules, Why Aren't They Building Fancy Molecular Machines?

Chemists can achieve great things, but have focused much of their effort on duplicating molecules found in nature and then making minor variants. As an example, take palytoxin, a molecule found in a Hawaiian coral. It was so difficult to make in the lab that it has been called "the Mount Everest of synthetic chemistry," and its synthesis was hailed as a triumph. Other efforts are poured into making small molecules with unusual bonding, or molecules of remarkable symmetry, like "cubane" and "dodecahedrane" (shaped like the Platonic solids they are named after).

Chemists, at least in the United States, regard themselves as natural scientists even when their life's work is the construction of molecules by artificial means.

Ordinarily, people who build things are called engineers. And indeed, at the University of Tokyo the Department of Synthetic Chemistry is part of the Faculty of Engineering; its chemists are designing molecular switches for storing computer data. Engineering achievements will require work directed at engineering goals.

Building Molecular Machines

Molecular engineers working toward nanotechnology need a set of molecular building blocks for making large, complex structures. Systematic building-block construction was pioneered by Bruce Merrifield, winner of the 1984 Nobel Prize in Chemistry. His approach, known as "solid phase synthesis," or simply "the Merrifield method," is used to synthesize the long chains of amino acids that form proteins. In the Merrifield method, cycles of chemical reaction each add one molecular building block to the end of a chain anchored to a solid support. This happens in parallel to each of trillions of identical chains, building up trillions of molecular objects with a particular sequence of building blocks. Chemists routinely use the Merrifield method to make molecules larger than palytoxin, and related techniques are used for making DNA in so-called gene machines: an ad from an Alabama company reads, "Custom DNA-Purified and Delivered in 48 hours."

While it's hard to predict how a natural protein chain will fold-they weren't designed to fold predictably-chemists could make building blocks that are larger, more diverse, and more inclined to fold up in a single, obvious, stable pattern. With a set of building blocks like these, and the Merrifield method to string them together, molecular engineers could design and build molecular machines with greater ease.

How do Researchers Design what they can't See?

To make a new molecule both its structure and the procedure to make it must be designed. Compared to gigantic science projects like the Superconducting Supercollider and the Hubble Space Telescope, working with molecules can be done on a shoestring budget. Still, the costs of trying many different procedures add up. To help predict in advance what will work and what won't, designers turn to models.

You may have played with molecular models in chemistry class: colored plastic balls and sticks that fit together like Tinker Toys. Each color represents a different kind of atom: carbon, hydrogen, and so on. Even simple plastic models can give you a feel for how many bonds each kind of atom makes, how long the bonds are, and at what angles they are made. A more sophisticated form of model uses only spheres and partial spheres, without sticks. These colorful, bumpy shapes are called CPK models, and are widely used by professional chemists. Nobel laureate Donald Cram remarks that "We have spent hundreds of hours building CPK models of potential complexes and grading them for desirability as research targets." His research, like that of fellow Nobelists Charles J. Pedersen and Jean-Marie Lehn, has focused on designing and making medium-sized molecules that self assemble.

Although physical models can't give a good description of how molecules bend and move, computer-based molecules can. Computer-based modeling is already playing a key role in molecular engineering. As John Walker (a founder and leader of Autodesk) has remarked, "Unlike all of the industrial revolutions that preceded it, molecular engineering requires, as an essential component, the ability to design, model, and simulate molecular structures using computers." This has not gone unnoticed in the business community. John Walker's remark was part of a talk on nanotechnology given at Autodesk, a leader in computer-aided design and one of the five largest software firms in the United States. Soon after this talk, the company made its first major investment in the computer-aided design of molecules.

MORE FAMILIAR KINDS OF ENGINEERING

Manufacturers and architects know that designs for new products and buildings are best done on a computer, by computer-aided design (CAD). The new molecular design software can be called molecular CAD, and in its forefront are researchers such as Jay Ponder of the Yale University Department of Molecular Biophysics and Biochemistry. Ponder explains that "There's a strong link between what molecular designers are doing and what architects do. Michael Ward of Du Pont is designing a set of building blocks for a Tinker Toy set so that you can build larger structures. That's exactly what we're doing with molecular modeling techniques.

"All the design and mechanical engineering principles that apply to building a skyscraper or a bridge apply to molecular architecture as well. If you're building a bridge, you're going to model it and see how many trucks can be on the bridge at the same time without it collapsing, what kind of forces you're going to apply to it, whether it can stand up to an earthquake.

"And the same process goes on in molecular design: You're designing pieces and then analyzing the stresses and forces and how they will change and perturb the structure. It's exactly the same as designing and building a building, or analyzing the stresses on any macroscale structure. I think it's important to get people to think in those terms.

"The molecular designer has to be creative in the same way that an architect has to be creative in designing a building. When people are looking at the interior of a protein structure and trying to redesign it to create a space that will have a particular function, such as binding to particular molecules, that's like designing a room to use as a dining room-one that will fit certain sizes of tables and certain numbers of guests. It's the same thing in both cases: You have to design a space for a function."

Ponder combines chemistry and computer science with an overall engineering approach: "I'm kind of a hybrid. I spend about half my time doing experiments and about half my time writing computer programs and doing computational work. In the laboratory, I create or design molecules to test some of the computational ideas. So I'm at the interface." The engineering perspective helps in thinking about where molecular research can lead: "Even though with nanotechnology we're at the nanometer scale, the structures are still big enough that an awful lot of things are classical. Again, it's really like building bridges-very small bridges. And so there are many almost standard mechanical-engineering techniques for architecture and building structures, such as stress analysis, that apply."

Getting to nanotechnology will require the work of experts in differing fields: chemists, who are learning how to make molecular machines; computer scientists, who are building the needed design tools; and perhaps STM and AFM experts, who can provide tools for molecular positioning. To make progress, however, these experts must do more than just work, they must work together. Because nanotechnology is inherently interdisciplinary, countries that draw hard lines between their academic disciplines, as the United States does, will find that their researchers have difficulty communicating and cooperating.

In chemistry today, a half-dozen researchers aided by a few tens of students and technicians is considered a large team. In aerospace engineering, enormous tasks like reaching the Moon or building a new airliner are broken down into tasks that are within the reach of small teams. All these small teams work together, forming a large team that may consist of thousands of engineers aided by many thousands of technicians. If chemistry is to move in the direction of molecular systems engineering, chemists will need to take at least a few steps in this direction.

In engineering, everyone knows that designing a rocket will require skills from many disciplines. Some engineers know structures, others know pumps, combustion, electronics, software, aerodynamics, control theory, and so on and so forth down a long list of disciplines. Engineering managers know how to bring different disciplines together to build systems.

In academic science, interdisciplinary work is productive and praised, but is relatively rare. Scientists don't need to cooperate to have their results fit together: they are all describing different parts of the same thing-nature-so in the long run, their results tend to come together into a single picture. Engineering, however, is different. Because it is more creative (it actually creates complex things), it demands more attention to teamwork. If the finished parts are going to work together, they must be developed by groups that share a common picture of what each part must accomplish. Engineers in different disciplines are forced to communicate; the challenge of management and team-building is to make that communication happen. This will apply to engineering molecular systems as much as it does to engineering computers, cars, aircraft, or factories.

Jay Ponder suggests that it's a question of perspective. "It's all a matter of what's perceived to be important by the different groups that have to come together to make this work: the chemists doing their bit and the computational people doing their bit. People have to come together and see the big picture. There are people who try to bridge the gaps, but they are rare compared to the people who just work in their own specialty." Progress toward nanotechnology will continue, and as it does, researchers trained as chemists, physicists, and the like will learn to talk to one another to solve new problems. They will either learn to think like engineers and work in teams, or they will be eclipsed by colleagues who do.

With all these problems, the advance toward nanotechnology steadily continues. Industry must gain ever-better control of matter to stay competitive in the world marketplace. The STM, protein engineering, and much of chemistry are driven by commercial imperatives. Focused efforts would yield faster advances, yet even without a clear focus, advances in this direction have an air of inevitability. As Bill DeGrado observes, "We really do have the tools. Experience has shown that when you have the analytic and synthetic tools to do things, in the end science goes ahead and does them-because they are doable."

Jay Ponder agrees: "Over the next few years, you're going to see slow evolutionary advances coming from people tinkering with molecular structures and figuring out their principles. People are going to work on a particular problem because they see some application for it or because they got grant funding for it. And in the process of doing something like improving a laundry detergent's ability to clean protein stains, Proctor and Gamble is going to help work out the principles for how to increase molecular stability, and to design spaces inside the molecules." For a variety of reasons, Japan's contribution to nanotechnology research promises to be excellent. While the United States has generally pursued researching this area with little sense of long-term direction, it appears that Japan has begun to take a more focused approach. Researchers there already have clear ideas about molecular machines-about what might work and what probably won't. Japanese researchers are accustomed to a higher level of interdisciplinary contact and engineering emphasis than are Americans. In the United States, we prize "basic science," often calling it "pure science," as if to imply that practical applications are a form of impurity. Japan instead emphasizes "basic technology."

Nanotechnology is a basic technology, and the Japanese recognize it as such. Recent changes at the Tokyo Institute of Technology-Japan's equivalent of MIT-reflect their views of promising directions for future research. For many decades, Tokyo Tech has had two major divisions: a Faculty of Science and a Faculty of Engineering. To these is now being added a Faculty of Bioscience and Biotechnology, to consist of four departments: a Department of Bioscience, a Department of Bioengineering, a Department of Biomolecular Engineering, and what is termed a "Department of Biostructure." The creation of a new faculty in a major Japanese university is a rare event. What U.S. university has a department explicitly devoted to molecular engineering? Japan has both the departments at Tokyo Tech and Kyoto University's recently established Department of Molecular Engineering.

Japan's Institute for Physical and Chemical Research (RIKEN) has broad-based interdisciplinary strength. Hiroyuki Sasabe, head of the Frontier Materials Research Program at RIKEN, notes that the institute has expertise in organic synthesis, protein engineering, and STM technology. Sasabe says that his laboratory may need a molecular manipulator of the sort described in the next chapter to accomplish its goals in molecular engineering. Research consortia in Japan are also moving toward nanotechnology. The Exploratory Research for Advanced Technology Organization (ERATO) sponsors many three-to-five year projects in parallel, each with a specific goal. Consider the work in progress:

- Yoshida Nanomechanism Project
- Hotani Molecular Dynamic Assembly Project
- Kunitake Molecular Architecture Project
- Nagayama Protein Array Project
- Aono Atomcraft Project

These focus on different aspects of gaining control over matter at the atomic level. The Nagayama Protein Array Project aims to use proteins as engineering materials to move toward making new molecular devices. The Aono Atomcraft Project does not involve nuclear power-as its translation might imply-but is instead an interdisciplinary effort to use an STM to arrange matter on the atomic scale.

At some point, work on nanotechnology must move beyond spin-offs from other fields and undertake the design and construction of molecular machinery. This shift from opportunistic science to organized engineering requires a change in attitude. In this, Japan leads the United States.

Molecular nanotechnology will emerge step by step. Major milestones, such as the engineering of proteins and the positioning of individual atoms, have already been passed. To get a sense of the likely pace of developments, we need to look at how various trends fit together.

Computer-based molecular-modeling tools are spawning computer-aided design tools. These will grow more capable. The underlying technology base-computer hardware-has for decades been improving in price and performance on a steeply rising curve, which is generally expected to continue for many years. These advances are quite independent of progress in molecular engineering, but they make molecular engineering easier, speeding advances. Computer models of molecular machines are beginning to appear, and these will whet the appetites of researchers.

Progress in engineering molecular machines, whether using proximal probes or self-assembly, will eventually achieve striking successes; the objectives of research in Japan will begin to draw serious attention; understanding of the long-term promise of molecular engineering will become more widespread.

Some combination of these developments will eventually lead to a serious, public appraisal of what these technologies can achieve-and then the world of opinion, funding, and research fashion will change. Before, advances will be steady but haphazard; afterward, advances will be driven with the energy that flows into major commercial, military, and medical research programs, because nanotechnology will be recognized as furthering major commercial, military, and medical goals. The timing of subsequent events depends largely on when this threshold of serious attention is reached.

In making time estimates, people are prone to assume that a large change must take a long time. Most do, but not all. Pocket calculators had a dramatic effect on the slide-rule industry: they replaced it. The speed of this change caught the slide rule moguls by surprise, but the pace of progress in electronics didn't slow down merely to suit their expectations. One can argue that nanotechnology will be developed fast: many countries and companies will be competing to get there first. They will be driven onward both by the immense expected benefits-in many areas, including medicine and the environment-as well as by potential military applications. That is a powerful combination of motives, and competition is a powerful accelerator.

A counterargument, though, suggests that development will be slow: anyone who has done anything of significance in the real world of technology-doing a scientific experiment, writing a computer program, bringing a new product to market-knows that these goals take longer than expected. Indeed, Hofstadter's Law states that projects take longer than expected, even when Hofstadter's Law is taken into account. This principle is a good guide for the short term, and for a single project.

The situation differs, though, when many different approaches are being explored by many different groups over a period of years. Most projects may take longer than expected, but with many teams trying many approaches, one approach may prove faster than expected. The winner of a race is always faster than the average runner. John Walker notes, "The remarkable thing about molecular engineering is that it looks like there are many different ways to get there and, at the moment, rapid progress is being made along every path-all at the same time."

Also, technology development is like a race run over an unmapped course. When the first runners reach the top of a hill, they may see a shortcut. A trailing runner may decide to crash off into the bushes, and stumble across a bicycle and a paved road. The progress of technology is seldom predictable because progress often reveals new directions. How close we are to goal depends on whether technological advances are a constant pace of accelerating. In this diagram, the dashed line represents the current level of technology, and the large dot in the upper right represents a goal such as nanotechnology. With a straight-line advance, it's easier to estimate how far away a goal is. With an accelerating advance, a goal can be reached with little warning.

So how can we estimate a date for the arrival of nanotechnology? It's safest to take a cautious approach: When anticipating benefits, assume it's far off; when preparing for potential problems, assume it's right around the corner. The old folk saying applies: Hope for the best, prepare for the worst. Any dates assigned to "far off" and "right around the corner" can be no better than educated guesses-molecular behavior can be calculated, but not technology timetables of this sort. With those caveats, we would estimate that general-purpose molecular assemblers will likely be developed in the early decades of the twenty-first century, perhaps in the first.

John Walker, whose technological foresight has led Autodesk from start-up to a dominant role in its industry, points out that not long ago, "Many visionaries intimately familiar with the developments of silicon technology still forecast it would take between twenty and fifty years before molecular engineering became a reality. This is well beyond the planning horizon of most companies. But recently, everything has begun to change." Based on the new developments, Walker places his bet: "Current progress suggests the revolution may happen within this decade, perhaps starting within five years.

NANOTECHNOLOGICAL THRESHOLD

In this chapter, we outline how emerging technologies can lead to nanotechnology. The actual path to nanotechnology-the one that history books will record-could emerge from any one of the research directions in physics, biochemistry, and chemistry recounted in the last chapter, or (more likely) from a combination of them. The availability of so many good options builds confidence that the goal can be reached, even while it decreases confidence that some particular path will be fastest. To see how advances might cross the gap from present technology to early nanotechnology, let's follow one path out of the many possible.

One way to bridge the gap would through the development of an AFM-based molecular manipulator capable of doing primitive molecular manufacturing. This device would combine a simple molecular device-a molecular gripper-with an AFM positioning mechanism. An AFM can move its tip with precision; a molecular manipulator would add a gripper to the tip to hold a molecular tool. A molecular manipulator of this kind would guide chemical reactions by positioning molecules, like a slow, simple, but enormous assembler. (In our standard simulation view, where a molecular assembler arm fits in a room, the AFM apparatus of a molecular manipulator would be the size of a moon.) Despite its limits, an AFM molecular manipulator will be a striking advance.

How might this advance occur? Since we're choosing one path out of many possible, we may as well include more details and tell a story. (A more technical description of a device like the following can be found in Nature; see the technical bibliography).

Developing a Molecular Manipulator

Several years ago, researchers at the University of Brobdingnag began work on developing a molecular manipulator. To reach this goal, a team of a dozen physicists, chemists, and protein researchers banded together (some working full time, some part time) and began the creative teamwork needed to solve the basic problems.

First they needed to attach a gripper to an AFM tip. As grippers, they chose fragments of antibody molecules, the selectively sticky proteins that the immune system uses to bind and identify germs. If they could get the "back" of the molecule stuck onto a tip, then the "front" could bind and hold molecular tools. (The advantage of antibody fragments was this: freedom of tool choice. Since the late 1980s, researchers had been able to generate antibodies able to bind almost any preselected molecule-or molecular tool.)

A molecular manipulator would bind and position reactive molecular tools to build up a workpiece, molecule by molecule. In parallel, the U. Brob AFM researchers worked on placing tips in a precise location and then holding them there with atom : accuracy for seconds at a time. This proved straightforv ard. They used techniques developed elsewhere during the early 1990s, adding only modest refinements.

They now had their gripper and a way of putting it where they wanted it, but they needed a set of tools. The gripper was like the chuck of a drill, waiting to have different bits fitted into its tool-holder slot. So as the final step, the synthetic chemists on the team made a dozen different molecular tools, all identical at one end but different at the other. The similar parts all bound to the same antibody tool-holder, slotting neatly into position. The different parts were all chemically reactive in different ways. Developing the molecular tool kit was the toughest part of the project; it took about as much work as had gone into duplicating the palytoxin molecule back in the

1980s. None of the tasks in the project demanded the solution of a deep scientific puzzle, and none demanded the solution of a notoriously difficult engineering problem. Each task had many possible solutions, the problem was to find a compatible set of solutions and apply them. After a few years, the solutions came together and the U. Brob research team began building new molecules by molecular manipulation. Now many teams are doing likewise.

Grippers and Tools

To build something with the U. Brob team's AFM-based molecular manipulator system, you use it as follows: First, choose a surface to build on and place it under the tip in a pool of liquid. Then dunk the AFM tip into the liquid, bringing it down to the surface, and back it off a little. Construction can now begin as soon as a tool is loaded into the gripper.

Tubes and pumps can flow different liquids over the surface and past the gripper, carrying different tool molecules. If you want to do something with a tool of Type A, you wash in the proper liquid. Once it is in the gripper, you can use the AFM mechanism to move it around and put it where you want it. Move it up to the surface at a convenient spot, wait a few seconds, and it reacts, forming a bond and leaving a molecular fragment attached to the spot you chose. To add a different fragment, you can use a tool of Type B: you back up the tip, flow in a fresh liquid carrying the new tools, and in a moment a tool of the new type is bound in place and ready to apply, either on or alongside the first spot. Step by step, you build up a precise molecular structure.

Each step takes only seconds. Molecular tools pop into the gripper in a fraction of a second, and used tools pop off at the same rate. Once the tip has positioned a molecule, it reacts quickly, about a million times faster than unwanted reactions at other sites. In this way, the molecular manipulator gives good control of where reactions will occur (though it is not as reliable as an advanced assembler would be). It is fairly fast by a chemist's standards-per cycle-but still a million times slower than an advanced assembler. It can perform a variety of steps, but isn't as flexible and capable as an advanced assembler. In short, it is hardly the last word in nanotechnology, yet is a great advance over what has gone before.

Products

With its ability to accelerate desired reactions by a factor of a million or so, the U. Brob team's molecular manipulator can perform 10,000 to 100,000 steps with good reliability. Back in the 1980s, chemists making protein molecules struggled to perform just one hundred steps. The U. Brob research team (and its many imitators) can now build structures that are stronger and easier to design than proteins: not floppy, folded chains, but rugged objects held together by a sturdy network of bonds. Though not as strong and dense as diamond, these structures are like bits of a tough engineering plastic. A specially adapted computer-aided design system makes it easy to design molecular objects made from these materials.

Yet the AFM-based molecular manipulator has one grave disadvantage: It does chemistry one molecule at a time, and it ties up a machine as expensive as a car for hours or days to produce that one large molecule. Some molecules, though, are valuable enough to be worth building even one at a time. These draw prompt attention.

A single molecule isn't much use as a dye, a drug, or a floor wax, but it can have substantial value if it provides useful information. The U. Brob team quickly publishes a pile of scientific papers based on experiments with single molecules: they build a molecule, probe it, report the results, and build another. Some of these results show chemists elsewhere in the multibillion-dollar chemical industry how to design new catalysts, molecules that can help make other molecules more cheaply, cleanly, and efficiently. This information is worth a lot.

Three new products of special interest are among the first to be made. The first-molecular electronics-begins with experiments conducted by a research group at a computer chip company. They use their molecular manipulator to build single molecules and probe them, gradually learning how to build the parts needed for molecular electronic computers. These new computers don't immediately become practical, because the costs are too high for making such large molecules with AFM-based technology. Yet some companies begin to produce simpler molecular electronic devices for use in sensors and specialized high-speed signal processing. A specialty industry is born and begins to expand.

The second product is a gene reader, a complex molecular device built on the surface of a chip. The biologists who built the reader combined proteins borrowed from cells with special-purpose molecular machines designed from scratch. The result was a molecular system that binds DNA molecules and pulls them past a read-head-like tape through a tape recorder. The device works as fast as some naturally occurring molecular machines that read DNA, with one key advantage: it outputs its data electronically. At that speed, a single device can read a human genome in about a year. Though still too expensive for a doctor's office, these readers are promptly in great demand from research laboratories. Another small industry is born.

The third product is far more important, in the long run: replacement tips for molecular manipulators, grippers, and tools that are better than the originals. With these new, more versatile devices, researchers are now building more ambitious products and tools.

NANOTECHNOLOGY: THE NEXT LEAP

While the physicist-led team at U. Brob was finishing its work on the AFM-based molecular manipulator, a chemist-led team at the University of Lilliput was working furiously. They saw the U. Brob desktop machine as too large and its expected products as too expensive. Even back in the 1980s, David Biegelsen of the Xerox Palo Alto Research Center had noted, "The main drawback I see to using a hybrid protoassembler [AFM-based molecular manipulator] is that it would take a long time to build just one unit. Building requires a series of atom-by-atom construction steps. It would be better to build in parallel from the very beginning, making many trillions of these molecules all at the same time. I think there is tremendous power in parallel assembly. Maybe another field, chemistry or biology, offers a better way to do it." The chemists at U. Lill aimed to develop that better way, building first simple and then more and more complex molecular machines. The eventual result was a primitive molecular assembler able to build molecular objects by the trillions.

Chemist's Tools

How did the chemists achieve this? During the years when the U. Brob team was developing the molecular manipulator, researchers working in protein science and synthetic chemistry had made better and better systems of molecular building blocks. Chemists were well prepared for doing this: by the late 1980s, it had become possible to build stable structures the size of medium-sized protein molecules, and work had begun to focus on making these molecules perform useful work by binding and modifying other molecules. Chemists learned to use these sophisticated catalysts-early molecular devices-to make their own work easier by helping in the manufacture of still more large molecules.

Another traditional chemist's tool was software for doing computer-aided design. The early software designed by Jay Ponder and Frederic Richards of Yale University ultimately led to semi-automatic tools for designing molecules of a particular shape and function. Chemists then could easily design molecules that would self-assemble into larger structures, several tens of nanometers across.

MOLECULAR CONSTRUCTION MACHINES

These advances in software and chemical synthesis let the U. Lill team tackle the task of building a primitive version of a molecular assembler. Although they couldn't build anything as complex as a nanocomputer or as stiff as diamond, they didn't need to. Their design used sliding molecular rods to position a molecular gripper much like the gripper used at U. Brob, again using the surrounding liquid to control which tool the gripper held. Instead of an AFM's electronic controls, they used the surrounding liquid to control the position of the rods as well. In a neutral solution, the rods would withdraw; in an acid solution, they would extend. How far they moved depended on what other molecules were around to lodge in special pockets and block the motion.

Their primitive assemblers built much the same sorts of products that the U. Brob molecular manipulator did; the tools were similar, and speed and accuracy were about the same. Yet there was one dramatic advantage: About 1,000,000,000,000,000,000,000 U. Lill assemblers could fit in the space occupied by one U. Brob manipulator, and it was easy to produce a mere 1,000,000,000,000,000 times as much product at the same cost. With the first, primitive assemblers, construction was slow because each step required new liquid baths and several seconds of soaking and waiting, and a typical product might take thousands of steps. Nonetheless, the U. Lill team made a lot of money licensing their technology to researchers trying to commercialize products they had first researched with the U. Brob machine. After starting an independent company (Nanofabricators, Inc.), they poured their research efforts into building better machines.

Within a few years, they had assemblers with multiple grippers, each loaded with a different kind of tool; flashes of colored light would flip molecules from state to state (they copied these molecules from the pigments of the retina of the eye); flipping molecules would change tools and change rod positions. Soaking and waiting become a thing of the past, and soon they were pouring out parts that, when mixed with liquid and added to dishes with special blank chips would build up the dense memory layers that made possible the Pocket Library. That was when things started moving fast. The semiconductor industry went the way of the vacuum tube industry. Money and talent poured into the new technology. Molecular CAD tools got better, assemblers made it easy to build what was designed, and fast production and testing made molecular engineering as easy as playing with software. Assemblers got better, faster, and cheaper. Researchers used assemblers to build nanocomputers, and nanocomputers to control better, faster assemblers. Using tools to build better tools is an ancient story. Within a decade, almost anything could be made by molecular manufacturing, and was.

NANOCOMPUTERS

While the first, primitive assemblers were controlled by changing what molecules are in the solution around the device, getting the speed and accuracy wanted for large-scale manufacturing takes real computation. Carl's setup uses a combination of special-purpose molecule processors and general-purpose assemblers, all controlled and orchestrated by nanocomputers.

Computers back in the 1990s used microelectronics. They worked by moving electrical charge back and forth through conducting paths-wires, in effect-using it to block and unblock the flow of charge in other paths. With nanotechnology, computers are built from molecular electronics. Like the computers of the 1990s, they use electronic signals to weave the patterns of digital logic. Being made of molecular components, though, they are built on a much smaller scale than 1990s computers, and work much faster and more efficiently. On the scale of our simulated molecular world, 1990s computer chips are like landscapes, while nanocomputers are like individual buildings. Carl's desktop PC contains over a trillion nanocomputers, enough to out-compute all the microelectronic computers of the twentieth century put together.

Back in the dark ages of the 1980s, an exploratory engineer proposed that nanocomputers could be mechanical, using sliding rods instead of moving electrons as shown in Figure 8. These molecular mechanical computers were much easier to design than molecular electronic computers would have been. They were a big help in getting some idea of what nanotechnology could do.

An electronic transistor lets current flow when a negative electric charge is applied and blocks current when a positive charge is applied. The mechanical "transistor" lets the horizontal rod move when the vertical rod is down, and blocks the horizontal rod when the vertical rod is up. Either device can be used to build logic gates and computers.

Even back then, it was pretty obvious that mechanical computers would be slower than electronic computers. Carl's molecular electronic PC would have been no great surprise, though nobody knew just how to design one. When nanotechnology actually arrived and people started competing to build the best possible computers, molecular electronics won the technology race. Still, mechanical nanocomputers could have done all the nanocomputing jobs at Desert Rose: ordinary, everyday molecular manufacturing just doesn't demand the last word in computer performance.

For Carl, the millions of nanocomputers in the milky waters of his building ponds are just extensions of machines on his desk, machines there to help him run his business and deliver products to his customers-or, in the case of the Red Cross emergency, to help provide time-critical emergency supplies. By reserving those three separate ponds, Carl can either build three different kinds of equipment for the Red Cross or use all the ponds to mass-produce the first thing on the Red Cross list: emergency shelters for ten thousand people. The software is ready, the plumbing is fine, the drums of building materials are all topped up, the Special Mix for this job is loaded: the build is ready to start. "Okay," Carl tells the computer, "build Red Cross tents." Computer talks to nanocomputers. In all three pools, nanocomputers talk to assemblers. The build begins.

ASSEMBLING The PRODUCTS

Some of the building done at Desert Rose Industries uses assemblers much like the ones we saw in the first hall of the plant tour, back in the simulated molecular world of the Silicon Valley Faire. As seen in simulation, they are big, slow, computer-controlled things moving molecular tools. With the right instructions and machinery to keep them supplied with molecular tools, these general-purpose assemblers can build almost anything. They're slow, though, and take a lot of energy to run. Some of the building uses special-purpose assembly systems in the molecule-processing style, like the systems in the basement we saw in the tour of a simulated molecular factory. The special-purpose systems are all moving belts and rollers, but no arms. This is faster and more efficient, but for quantity orders, cooling requirements limit the speed.

It's faster to use larger, prefabricated building blocks. Desert Rose uses these for most of their work, and especially for rush orders like the one Carl just set up. Their underground warehouse has room-sized bins containing upward of a thousand tons of the most popular building blocks, things like structural fibers. They're made at plants on the West Coast and shipped here by subway for ready use. Other kinds are made on site using the special-purpose assemblers. Carl's main room has several cabinet-sized boxes hooked up to the plumbing, each taking in raw materials, running them through this sort of specialized molecular machinery, and pumping out a milky syrup of product. One syrup contains motors, another one contains computers, and another is full of microscopic plug-in light sources. All go into tanks for later use.

Now they're being used. The mix for the Red Cross tent job is mostly structural fiber stronger than the old bulletproof-vest materials. Other building blocks also go in, including motors, computers, and dozens of little struts, angle brackets, and doohickies. The mix would look like someone had stirred together the parts from a dozen toy sets, if the parts were big enough to see. In fact, though, the largest parts would be no more than blurry dots, if you saw one under a normal optical microscope.

The mix also contains block-assemblers, floating free like everything else. These machines are big, about like an office building in our simulation view with the standard settings. Each has several jointed arms, a computer, and several plugs and sockets. These do the actual construction work.

To begin the build, pumps pour the mix into a manufacturing pond. The constant tumbling motions of microscopic things in liquids would be too disorganized for building anything so large as a tent, so the block-assemblers start grabbing their neigbours. Within moments, they have linked up to form a framework spread through the liquid. Now that they are plugged together, they divide up jobs, and get to work. Instructions pour in from Carl's desktop computer.

The block-assemblers use sticky grippers to pull specific kinds of building blocks out of the liquid. They use their arms to plug them together. For a permanent job, they would be using blocks that bond together chemically and permanently. For these temporary tents, though, the Red Cross design uses a set of standard blocks that are put together with amazingly ordinary fasteners: these blocks have snaps, plugs, and screws, though of course the parts are atomically perfect and the threads on the screws are single helical rows of atoms. The resulting joints weaken the tent's structure somewhat, but who cares? The basic materials are almost a hundred times stronger than steel, so there is strength to waste if it makes manufacturing more convenient.

Fiber segments snap together to make fabrics. Some segments contain motors and computers, linked by fibers that contain power and data cables. Struts snap together with more motors and computers to make the tent's main structures. Special surfaces are made of special building blocks. From the human perspective, each tent is a lightweight structure that contains most of the conveniences and comforts of an apartment: cooking facilities, a bathroom, beds, windows, air conditioning, specially modified to meet the environmental demands of the quake-stricken country. From a builder's perspective, especially from a nanomachine's point of view, the tent is just structure slapped together from a few hundred kinds of prefab parts.

In a matter of seconds, each block-assembler has put together a few thousand parts, and its section of the tent is done. In fact, the whole thing is done: many trillions of hands make light work. A crane swings out over the pond and starts plucking out tent packages as fresh mix flows in.

Maria's concern has drawn her back to the plant to see how the build is going. "It's coming along," Carl reassures her. "Look, the first batch of tents is out." In the warehouse, the first pallet is already stacked with five layers of dove-gray "suitcases": tents dried and packed for transport. Carl grabs a tent by the handle and lugs it out the door. He pushes a tab on the corner labeled "Open," and it takes over a minute to unfold to a structure a half-dozen paces on a side. The tent is big, and light enough to blow away if it didn't cling to the ground so tightly. Maria and Carl tour the tent, testing the appliances, checking the construction of furniture: everything is extremely lightweight compared to the bulk-manufactured goods of the 1990s, tough but almost hollow.

Like the other structures, the walls and floors are full of tiny motors and struts controlled by simple computers like the ones used in twentieth-century cars, televisions, and pinball machines. They can unfold and refold. They can also flex to produce sound like a high-quality speaker, or to absorb sound to silence outdoors racket. The whole three-room setup is small and efficient, looking like a cross between a boat cabin and a Japanese business hotel room. Outside, though, it is little more than a box. Maria shakes her head, knowing full well what architects can do these days when they try to make a building really fit its site. Oh well, she thinks, These won't be used for long.

Behind the Scenes and Afterward

Desert Rose Industries and other manufacturers can make almost anything quickly and at low cost. That includes the tunneling machines and other equipment that made the subway system they use for shipping. Digging a tunnel from coast to coast now costs less than digging a single block under New York City used to. It wasn't expensive to get a deep-transit terminal installed in their basement. Just as the tents aren't mere bundles of canvas, these subways aren't slow things full of screeching, jolting metal boxes. They're magnetically levitated to reach aircraft speeds-as experimental Japanese trains were in the late 1980s-making it easy for Carl and Maria to give their customers quick service. There's still a road leading to the plant, but nobody's driven a truck over it for years.

They only take in materials that they will eventually ship out in products, so there's nothing left over, and no wastes to dump. One corner of the plant is full of recycling equipment. There are always some obsolete parts to get rid of, or things that have been damaged and need to be reworked. These get broken down into simpler molecules and put back together again to make new parts.

The gunk in the manufacturing ponds is water mixed with particles much finer than silt. The particles-fasteners, computers, and the rest-stay in suspension because they are wrapped in molecular jackets that keep them there. This uses the same principle as detergent molecules, which coat particles of oily dirt to float them away.

Though it wouldn't be nutritious or appetizing, you could drink the tent mix and be no worse for it. To your body, the parts and their jackets, and even the nanomachines, would be like so many bits of grit and sawdust. (Grandma would have called it roughage.)

Carl and Maria get their power from solar cells in the road, which is the only reason they bothered having it paved. In back of their plant stands what looks like a fat smokestack. All it produces, though, is an updraft of clean, warm air. The darkly paved road, baking in the New Mexico sun, is cooler than you might expect: it soaks up solar energy and makes electricity, instead of just heat. Once the power is used, it turns back into heat, which has to go somewhere. So the heat rises from their cooling tower instead of the road, and the energy does useful work on the way.

Some products, like rocket engines, are made more slowly and in a single piece. This makes them stronger and more permanent. The tents, though, don't need to be superstrong and are just for temporary use. A few days after the tents go up, the earthquake victims start to move out into new housing (permanent, better-looking, and very earthquake resistant). The tents get folded and shipped off for recycling.

Recycling things built this way is simple and efficient: nanomachines just unscrew and unsnap the connectors and sort the parts into bins again. The shipments Desert Rose gets are mostly recycled to begin with. There's no special labeling for recycled materials, because the molecular parts are the same either way.

For convenience (and to keep the plant small), Carl and Maria get most of their parts prefabricated, even though they can make almost anything. They can even make more production equipment. In one of their manufacturing ponds, they can put together a new cabinet full of special-purpose assemblers. They do this when they want to make a new type of part in-house. Like parts, the part-assemblers are made by special-purpose assemblers. Carl can even make big vats in medium-size vats, unfolding them like tents.

If Desert Rose Industries needed to double capacity, Carl and Maria could do it in just a few days. They did this once for a special order of stadium sections. Maria got Carl to recycle the new building before its shadow hurt their cactus garden.

In the Desert Rose Industries scenario, manufacturing has become cheap, fast, clean, and efficient. Using fast, precise machines to handle matter in molecular pieces makes it easy for nanotechnology to be fast, clean, and efficient. But for it to be cheap, the manufacturing equipment has to be cheap.

The Desert Rose scenario shows how this can work. Molecular-manufacturing equipment can be used to make all the parts needed to build more molecular manufacturing equipment. It can even build the machines needed to put the parts together. This resembles an idea developed by NASA for a self-expanding manufacturing complex on the Moon, but made faster and simpler using molecular machines and parts.

In the early days of nanotechnology, there won't be as many different kinds of machines as there are at Desert Rose. One way to build a lot of molecular manufacturing equipment in a reasonable time would be to make a machine that can be used to make a copy of itself, starting with special but simple chemicals. A machine able to do this is called a "replicator." With a replicator and a pot full of the right fuel and raw materials, you could start with one machine, then have two, four, eight, and so on.

This doubling process soon makes enough machines to be useful. The replicators-each including a computer to control it and a general-purpose assembler to build things-could then be used to make something else, like the tons of specialized machines needed to set up a Desert Rose manufacturing plant. At that point, the replicators could be discarded in favour of those more efficient machines.

Replicators are worth a closer look, though, because they show how quickly molecular manufacturing systems can be used to build more manufacturing equipment. If we were in one of our standard simulation views, the submicroscopic device at the top of the picture would be like a huge tank, three stories tall when lying on its side. Most of its interior is taken up by a tape memory system that tells how to move the arm to build all the parts of the replicator, except the tape itself. The tape gets made by a special tape-copying machine. At the right-hand end of the replicator are pores for bringing in fuel and raw-material molecules, and machinery for processing them. In the middle are computer-controlled arms, like the ones we saw on the plant trip. These do most of the actual construction.

The steps in the cycle-using a copy to block the tube, beginning a fresh copy, then releasing the old one-illustrate one way for a machine to build a copy of itself while floating in a liquid, yet doing all its construction work inside, in vacuum. (It's easier to design for vacuum, and this is exploratory-engineering work, so easier design is better design.) Calculations suggest that the whole construction cycle can be completed in less than a quarter hour, since the replicator contains about a billion atoms, and each arm can handle about a million atoms per second. At that rate, one device can double and double again to make trillions in about ten hours.

Each replicator just sits in a chemical bath, soaking up what it needs and making more replicators. Eventually, either the special chemicals run out or other chemicals are added to signal them to do something else. At that point, they can be reprogrammed to produce anything else you please, so long as it can be extruded from the front. The products can be long, and can unfold or be pieced together to make larger objects, so the size of these initial replicators-smaller than a bacterium-would be only a temporary limitation.

From the molecular manipulators and primitive assemblers described in the last chapter, the most likely path to nanotechnology leads to assemblers with more and more general capabilities. Still, efficiency favours special-purpose machines, and the Desert Rose scenario didn't make much use of general assemblers. Why bother making general-purpose assemblers in the first place?

To see the answer, turn the question around and ask, Why not build such a tool? There is nothing outstandingly difficult about a general assembler, as molecular machinery goes. It will just be a device with good, flexible positional control and a system to feed it a variety of molecular tools. This is a useful, basic capability. General-purpose assemblers could always be replaced by a lot of specialized devices, but to build those specialized devices in the first place, it makes sense to come up with a more flexible, general-purpose system that can just be reprogrammed.

So, general purpose machines are likely to find use in making short production runs of more specialized devices. Ralph Merkle, a computers and security expert at Xerox Palo Alto Research Center, sees this as paralleling the way manufacturing works today: "General purpose devices could do many tasks, but they'll do them inefficiently. For any given task, there will be one or a few best ways of doing it, and one or a few special purpose devices that are finely tuned to do that one task. Nails aren't made by a general-purpose machine shop, they're made by nail-making machines. Making nails with a general-purpose machine shop would be more expensive, more difficult, and more time-consuming. Likewise, in the future we won't see a proliferation of general-purpose self-replicating systems, we'll see specialization for almost every task."We've surveyed a lot of devices: assemblers of various flavors, nanocomputers, disassemblers, replicators, and others. What's important about these is not the exact distinctions between them, but the capabilities that they will give and the effects they will have on human lives. Again, we are suspending discussion of potential misapplications until later.

If we tease apart the implications of what we've seen in the Desert Rose scenario, we can analyze some of the key impacts of molecular manufacturing in industry, science, and medicine.

THE INDUSTRY AND TECHNOLOGY

At its base, nanotechnology is about molecular manufacturing, and manufacturing is the basis of much of today's industry. This is why Desert Rose made a good starting point for describing the possibilities of a nanotechnological world. From an industrial perspective, it makes sense to think of nanotechnology in terms of products and production.

New Products: Today, we handle matter crudely, but nanotechnology will bring thorough control of the structure of matter, the ability to build objects to atom-by-atom specifications. This means being able to make almost anything. By comparison, even today's range of products will feel very limited. Nanotechnology will make possible a huge range of new products, a range we can't envision today. Still, to get a feel for what is possible, we can look at some easily imagined applications.

Reliable Products: Today, products often fail, but for failures to occur-for a wing to fall off an airplane, or a bearing to wear out-a lot of atoms have to be out of place. In the future, we can do better. There are two basic reasons for this: better materials and better quality control, both achieved by molecular manufacturing. By using materials tens of times stronger than steel, as Desert Rose did, it will be easy to make things that are very strong, with a huge safety margin. By building things with atom-by-atom control, flaws can be made very rare and extremely small-nonexistent, by present standards.

With nanotechnology, we can design in big safety margins and then manufacture the design with near-perfection. The result will be products that are tough and reliable. (There will still be room for bad designs, and for people who wish to take risks in machines that balance on the edge of disaster.)

Intelligent Products: Today, we make most things from big chunks of metal, wood, plastic, and the like, or from tangles of fibers. Objects made with molecular manufacturing can contain trillions of microscopic motors and computers, forming parts that work together to do something useful. A climber's rope can be made of fibers that slide around and reweave to eliminate frayed spots. Tents can be made of parts that slide and lock to turn a package into a building. Walls and furniture can be made to repair themselves, instead of passively deteriorating.

On a mundane level, this sort of flexibility will increase reliability and durability. Beyond this, it will make possible new products with abilities we never imagined we needed so badly. And beyond even this, it will open new possibilities for art.

***Inexpensive Production*:** Today, production requires a lot of labor, either for making things or for building and maintaining machines that make things. Labor is expensive, and expensive machines make automation expensive, too. In the Desert Rose scenario, we got a glimpse of how molecular manufacturing can make production far less expensive than it is today. This is perhaps the most surprising conclusion about nanotechnology, so we'll take a closer look at it in the next chapter.

***Clean Production*:** Today, our manufacturing processes handle matter sloppily, producing pollution. One step puts stuff where it shouldn't be; the next washes it off the product and into the water supply. Our transportation system worsens the problem as unreliable trucks and tankers spill noxious chemicals over the land and sea. Everything is expensive, so companies skimp on even the half-effective pollution controls that we know how to build.

Nanotechnology will mean greater control of matter, making it easy to avoid pollution. This means that a little public pressure will go a long way toward a cleaner environment. Likewise, it will make it easy to increase efficiency and reduce resource requirements. Products, like the Red Cross tents at Desert Rose, can be made of snap-together, easily recyclable parts. Sophisticated products could even be made from biodegradable materials. Nanotechnology will make it easy to attack the causes of pollution at their technological root.

Nanotechnology will have great applications in the field of industry, much as transistors had great applications in the field of vacuum tube electronics, and democracy had great applications in the field of monarchy. It will not so much advance twentieth-century industry as replace it-not all at once, but during a thin slice of historical time.

Chemistry: Today, chemists work with huge number of molecules and study them using clever, indirect techniques. Making a new molecule can be a major project, and studying it can be another. Molecular manufacturing will help chemists make what they want to study, and it will help them make the tools they need to study it. Nanoinstruments will be used to prod, measure, and modify molecules in a host of ways, studying their structures, behaviors, and interactions.

Materials: Today, materials scientists make new superconductors, semiconductors, and structural materials by mixing and crushing and baking and freezing, and so forth. They dream of far more structures than they can make, and they stumble across more things than they plan. With molecular manufacturing, materials science can be much more systematic and thorough. New ideas can be tested because new materials can be built according to plan (rather than playing around, groping for a recipe).

This need not rule out unexpected discoveries, since experiments-even blind searches-will go much faster. A few tons of raw materials would be enough to make a billion samples, each a cubic micron in size. In all of history so far, materials scientists have never tested so many materials. With nanoinstruments and nanocomputers, they could. One laboratory could then do more than all of today's materials scientists put together.

Biology: Today, biologists use a host of molecular devices borrowed from biology to study biology. Many of these can be viewed as molecular machines. Nanotechnology will greatly advance biology by providing better molecular devices, better nanoinstruments. Some cells have already been mapped in amazing molecular detail, but biology still has far to go. With nanoinstruments (including molecule-by-molecule disassemblers), biologists will at last be able to map cells completely and study their interactions in detail. It will become easy not only to find molecules in cells, but to learn what they do. This will help in understanding disease and the molecular requirements for health, enormously advancing medicine.

Computation: Today, computers range from a million to a billion times faster than an old desktop adding machine, and the results have been revolutionary for science. Every year, more questions can be answered by calculations based on known principles of physics. The advent of nanocomputers-even slow, miserable, mechanical nanocomputers-will give us practical machines with a trillion times the power of today's computers (essentially by letting us package a trillion computers in a small space, without gobbling too much moncy or energy.) The consequences will again be revolutionary.

Physics: The known principles of physics are adequate for understanding molecules, materials, and cells, but not for understanding phenomena on a scale that would still be submicroscopic if atoms were the size of marbles. Nanotechnology can't help here directly, but it can provide manufacturing facilities that will make huge particle accelerators economical, where today they strain national budgets.

More generally, nanotechnology will help science wherever precision and fine details are important. Science frequently proceeds by trying small variations in almost identical experiments, comparing the results. This will be easier when molecular manufacturing can make two objects that are identical, molecule by molecule. In some areas, today's techniques are not only crude, but destructive. Archaeological sites are unique records of the human past, but today's techniques throw away most information during the dig, by accident. Future archaeologists, able to sift soil not speck by speck but molecule by molecule, will be grateful indeed to those archaeologists who today leave some ground undisturbed.

Of all the areas where the ability to manufacture new tools is important, medicine is perhaps the greatest. The human body is intricate, and that intricacy extends beyond the range of human vision, beyond microscopic imaging, down to the molecular scale. "Molecular medicine" is an increasingly popular term today, but medicine today has only the simplest of molecular tools. As biology uses nanoinstruments to learn about disease and health, we will learn the physical requirements for restoring and maintaining health. And with this knowledge will come the tools needed to satisfy those requirements-tools ranging from improved pharmaceuticals to devices able to repair cells and tissues through molecular surgery.

Advanced medicine will be among the most complex and difficult applications of nanotechnology. It will require great knowledge, but nanoinstruments will help gather this knowledge. It will pose great engineering challenges, but computers of trillionfold greater power will help meet those challenges. It will solve medical problems on which we spend billions of dollars today, in hopes of modest improvements.

Today, modern medicine often means an expensive way to prolong misery. Will nanomedicine be more of the same? Any reader over the age of, say, thirty knows how things start to go wrong: an ache here, a wrinkle there, the loss of an ability. Over the decades, the physical quality of life declines faster and faster-the limits of what the body can do become stricter-until the limits are those of a hospital bed. The healing abilities we have when young seem to fade away. Modern medical practice expends the bulk of its effort on such things as intensive care units, dragging out the last few years of life without restoring health.

Truly advanced medicine will be able to restore and supplement the youthful ability to heal. Its cost will depend on the cost of producing things more intricate than any we have seen before, the cost of producing computers, sensors, and the like by the trillions. To understand the prospects for medicine, like those for science and industry, we need to take a closer look at the cost of molecular manufacturing.

In earlier chapters, we have stepped forward and backward through time. The last step was a big one, leaping from small laboratory devices to the high-capacity industrial facility of the Desert Rose scenario. Our narrative crossed this gap in a single leap, but the world won't. To understand how nanotechnology might unfold, it makes sense to look at some of its easier and more difficult applications. The result won't be a timetable, or even a series of milestones, but it should give a better picture of what we can expect as nanotechnology develops from simple, crude, costly beginnings to a state of greater sophistication and lower cost.

Molecular manufacturing will make better products possible. We're likely to see some early applications in at least two areas: stronger materials and faster computers. Strong materials are simple, and will be hard to pass up. Computers are more complex, but the payoff will be enormous.

STRONG and LIGHTWEIGHT STRUCTURES

At the opposite extreme from molecular electronics-complex and at first worth billions of dollars per gram-are structural materials: worth only dollars per kilogram in most applications, but much simpler in structure. Once molecular manufacturing becomes inexpensive, structural materials will be important products.

These materials play a central role in almost everything around us, from cars and aircraft to furniture and houses. All of these objects get their size, shape, and strength from a structural skeleton of some sort. This makes structural materials a natural place to begin in understanding how nanotechnology can improve products.

Cars today are mostly made of steel, aircraft of aluminum, and buildings and furniture largely of steel and wood. These materials have a certain "strength-to-weight ratio" (more properly, a strength-to-density ratio). To make cars stronger, they'd have to be heavier; to make them lighter, they'd have to be weaker. Clever design can change this relationship a little, but to change it a lot requires a change of materials.

Making something heavy is easy: just leave a hollow space, then fill it with water, sand, or lead shot. Making something light and strong is harder, but often important. Automakers try to make cars lightweight, aircraft manufacturers try harder, and with spacecraft manufacturers it is an obsession. Reducing mass saves materials and energy.

The strongest materials in use today are mostly made of carbon. Kevlar, used in racing sails and bulletproof vests, is made of carbon-rich molecular fibers. Expensive graphite composites, used in tennis rackets and jet aircraft, are made using pure-carbon fibers. Perfect fibers of carbon-both graphite and diamond-would be even better, but can't be made with today's technology. Once molecular manufacturing gets rolling, though, such materials will be commonplace and inexpensive.

What will these materials be like? To picture them, a good place to start is wood. The structure of wood can vary from extremely light and porous, like balsa wood, to denser structures like oak. Wood is made by molecular machinery in plants from carbon-rich polymers, mostly cellulose. Molecular manufacturing will be able to make materials like these, but with a strength-to-weight ratio about a hundred times that of mediocre steel, and tens of times better than the best steel. Instead of being made of cellulose, these materials will be made of carbon in forms like diamond. Diamond is emphasized here not because it is shiny and expensive, but because it is strong and potentially cheap. Diamond is just carbon with properly arranged atoms. Companies are already learning to make it from natural gas at low pressure. Molecular manufacturing will be able to make complex objects of the stuff, built lighter than balsa wood but stronger than steel.

Products made of such materials could be startling by our present standards. Objects could be made that are identical in size and shape to those we make today, but simultaneously stronger and 90 percent lighter. This is something to keep in mind next time you're lugging a heavy object around. (If something needs weight to hold it in place, it would be more convenient to add this ballast when the thing is in its proper location than to build in the extra weight permanently.) Better structural materials will make aircraft lighter, stronger, and more efficient, but will have the greatest effect on spacecraft. Today, spacecraft can barely reach orbit with both a safety margin and a cargo. To get there at all, they have to drop off parts like boosters and tanks along the way, shedding weight.

Fast Development

In some areas of high technology-spaceflight has been a notorious example-it takes years, even decades, to try a new idea. This makes progress slow to a crawl. In other areas-software has been a shining example-new ideas can be tested in minutes or hours. Since the Space Shuttle design was frozen, personal computer software has come into existence and gone through several generations of commercial development, each with many cycles of building and testing.

Fast, Inexpensive Testing

Even in the days of the first operational molecular manipulators, experimentation is likely to be reasonably fast. Individual chemical steps can take seconds or less. Complex molecular objects could be built in a matter of hours. This will let new ideas be put into practice almost as fast as they can be designed.

Later assemblers will be even faster. At a millionth of a second per step, they will approach the speed of computers. And, as nanotechnology matures, experimenters will have more and more molecular instruments available to help them find out whether their devices work or not. Fast construction and fast testing will encourage fast progress.

At this point, the cost of materials and equipment for experiments will be trivial. No one today can afford to build Moon rockets on a hobby budget, but they can afford to build software, and many useful programs have been the result. There is no economic reason why nanomachines couldn't eventually be built with a hobby-size budget, though there are reasons-to be discussed in later chapters-for wanting to place limits on what can be built.

Prior Simplicity

Finally, established technologies are always pushing up against some limit; the easy opportunities have generally been exploited. In many fields, the limits are those of the properties of the materials used and the cost and precision of manufacturing. This is true for computers, for spacecraft, for cars, blenders, and shoes. For software, the limits are those of computer capacity and of sheer complexity (which is to say, of human intelligence). After molecular manufacturing develops certain basic abilities, a whole set of limits will fall, and a whole range of developments will become possible. Limits set by materials properties, and by the cost and precision of manufacturing, will be pushed way back. Competition, easy opportunities, and fast, low-cost experimentation should combine to yield an explosion of new products.

This does not mean immediately, and it does not apply to all imaginable nanotechnologies. Some technologies are imaginable and clearly feasible, yet dauntingly complex. Still, the above considerations suggest that a wide range of advances could happen at a brisk pace. The main bottleneck might seem to be a shortage of knowledgeable designers-hardly anyone knows both chemistry and mechanical design-but improving computer simulations will help. These simulations will let engineers tinker with molecular-machinery designs, absorbing knowledge of chemical rules without learning chemistry in the usual sense.

Climbing Complexity

Making familiar products from improved materials will increase their safety, performance, and usefulness.

It will also present the simplest engineering task. A greater change, though, will result from unfamiliar products made possible by new manufacturing methods. In talking about unfamiliar products, a hard-to-answer question arises: What will people want?

Products are typically made because their recipients want them. In our discussions here, if we describe something that people won't want, then it probably won't get built, and if it does get built, it will soon disappear. (The exceptions-fraud, coercion, persistent mistakes-are important, but in other contexts.) To anchor our discussion, it makes sense to look not at totally new products, but instead at new features for old products, or new ways to provide old services. This approach won't cover more than a fraction of what is possible, but will start from something sensible and provide a springboard for the imagination. As usual, we are describing possibilities, not making predictions. The possibilities focused on here arise from more complex applications of molecular manufacturing-nanotechnological products that contain nanomachines when they are finished. Earlier, we discussed strong materials. Now, we discuss some smart materials.

The goal of making materials and objects smart isn't new: researchers are already struggling to build structures that can sense internal and environmental conditions and adapt themselves appropriately. There is even a Journal of Intelligent Material Systems and Structures. By using materials that can adapt their shapes, sometimes hooked up to sensors and computers, engineers are starting to make objects they call "smart." These are the early ancestors of the smart materials that molecular manufacturing will make possible.

Today, we are used to having machines with a few visible moving parts. In cars, the wheels go around, the windshield wipers go back and forth, the antenna may go up and down, the seat belts, mirrors, and steering wheel may be motor-driven. Electric motors are fairly small, fairly inexpensive, and fairly reliable, so they are fairly common. The result is machines that are fairly smart and flexible, in a clumsy, expensive way.

In the Desert Rose scenario, we saw "tents" being assembled from trillions of submicroscopically small parts, including motors, computers, fibers, and struts. To the naked eye, materials made from these parts could seem as smooth and uniform as a piece of plastic, or as richly textured as wood or cloth-it is all a matter of the arrangement and appearance of the submicroscopic parts. These motors and other parts cost less than a trillionth of a dollar apiece. They can be quite reliable, and good design can make systems work smoothly even if 10 percent of a trillion motors burn out. Likewise for motor-controlling computers and the rest. The resulting machines can be very smart and flexible, compared to those of today, and inexpensive, too. When materials can be full of motors and controllers, whole chunks of material can be made flexible and controllable. The applications should be broad.

Surfaces surround us, and human-made surfaces-walls, roofs, and pavement-cover huge areas that matter to people. How can smart materials make a difference here?

The revolution in technology has come and gone, and you want to repaint your walls. Breathing toxic solvents and polluting water by washing brushes have passed into history, because paint has been replaced with smarter stuff. The mid-twentieth century had seen considerable progress in paints, especially the development of liquids that weren't quite liquid-they would spread with a brush, but didn't (stupidly) run and drip under their own weight.

This was an improvement, but the new material, "paperpaint," is even more cooperative.

Paperpaint comes in a box with a special trowel and pen. The paperpaint itself is a dry block that feels a lot like a block of wood. Following the instructions, you use the pen to draw a line around the edge of the area you want to paint, putting an X in the middle to show where you want the paint to go on; the line is made of nontoxic disappearing ink, so you can slop it around without staining anything. Using the trowel, you slice off a hunk of paperpaint-which is easy, because it parts like soft butter to the trowel, even though it behaves like a solid to everything else. Very high IQ stuff, that.

Next, you press the hunk against the X and start smoothing it out with the trowel. Each stroke spreads a wide swath of paperpaint, much wider than the trowel, but always staying within the inked line. A few swipes spreads it precisely to the edges, whereupon it smooths out into a uniform layer. Why doesn't it just spread itself? Experience showed that customers didn't mind the effort of making a few swipes and preferred the added control. The paperpaint consists of a huge number of nanomachines with little wheels for rolling over one another and little sticky pads for clinging to surfaces. Each has a simple, stupid computer on board. Each can signal its neigbours.

The whole mass of them clings together like an ordinary solid, but they can slip and slide in a controlled way when signaled. When you smooth the trowel over them, this contact tells them to get moving and spread out. When they hit the line, this tells them to stop. If they don't hit a line, they go a few handbreadths, then stop anyway until you trowel them again. When they encounter a line on all sides, word gets around, and they jostle around to form a smooth, uniform layer. Any that get scraped off are just so much loose dust, but they stick together quite well. This paint-stuff doesn't get anything wet, doesn't stain, and clings to surfaces just tightly enough to keep it from peeling off accidentally. If some experimentally minded child starts digging with a stick, makes a tear, and peels some off, it can be smoothed back again and will rejoin as good as new. The child may eat a piece, but careful regulation and testing has ensured that this is no worse than eating plain paper, and safer than eating a colorful Sunday newspaper page.

Many refinements are possible. Swipes and pats of the trowel could make areas thicken or thin, or bridge small holes (no more Spackling!). With sufficiently smart paperpaint, and some way to indicate what it should do, you can have your choice of textures. Any good design will be washable, and a better design would shed dirt automatically using microscopic brushes. Removal, of course, is easy: either you rip and peel (no scraping needed), or find that trowel, set the dial on the handle to "strip," and poke the surface a few times. Either way, you end up with a lump ready to pitch into the recycling bin and the same old wall you started with, bared to sight again.

Looking further at the human environment we find a lot of cloth and related materials, such as carpeting and shoes. The textile industry was at the cutting edge of the first industrial revolution, and the next industrial revolution will have its effects on textiles.

With nanotechnology, even the finest textile fibers could have sensors, computers, and motors in their core at little extra cost. Fabrics could include sensors able to detect light, heat, pressure, moisture, stress, and wear, networks of simple computers to integrate this data, and motors and other nanomechanisms to respond to it. Ordinary, everyday things like fabric and padding could be made responsive to a person's needs-changing shape, color, texture, fit, and so forth-with the weather and a person's posture or situation. This process could be slow, or it could be fast enough to respond to a gesture. One result would be genuine one-size-fits-all clothing (give or take child sizes), perfectly tailored off the rack, warm in winter, cool and dry in summer; in short, nanotechnology could provide what advertisers have only promised. Even bogus advertising gives a clue to human desires.

Throughout history, the human race has pursued the quest for comfortable shoes. With fully adjustable materials, the seemingly impossible goal of having shoes that both look good and feel good should finally be achieved. Shoes could keep your feet dry, and warm except in the Arctic, cool except in the tropics, and as comfortable as they can be with a person stepping on them.

Adaptive structures will be useful in furniture. Today, we have furniture that adapts to the human body, but it does so in an awkward and incomplete manner. It adapts because people grab cushions and move them around. Or a chair adapts because it is a hinged contraption that grudgingly bends and extends in a few places to suit a small range of preferred positions. Occasionally, one sees furniture that allegedly gives a massage, but in fact only vibrates.

These limitations are consequences of the expense, bulkiness, clumsiness, and unreliability of such things as moving parts, motors, sensors, and computers today. With molecular manufacturing, it will be easy to make furniture from smart materials that can adapt to an individual human body, and to a person's changing position, to consistently give comfortable support. Smart cushions could also do a better job of responding to hints in the form of pats, tugs, and punches. As for massage-a piece of furniture, no matter how advanced, is not the same as a masseuse. Still, a typical massage setting on a smart chair would not mean today's "vibrate medium vigorously," but something closer to "five minutes of shiatsu."

This tour through of the potential of smart matter has shown how we could get walls that look and sound as we wish, clothing, shoes, and furniture of greater comfort, and clean solar power. As one might expect, this just scratches the surface.

If you care to think of further applications, here are some ground rules: Components made by molecular manufacturing can be many tens of times stronger than steel, but materials made by plugging many components together will be weaker. For these, strengths in the range of cotton candy to steel seem achievable. The components will be sensitive to heat, and at high temperatures they will break down or burn. Many materials will be able to survive the temperature of boiling water, but only specialized designs would be oven-safe.

Color, texture, and (usually) sound should be controllable. Surfaces can be smooth and tightly sealed (this takes some cleverness). Motions can be fairly fast. Power has to come from somewhere; good sources include electricity, stored chemical energy, and light. If nanomachines or smart materials are dunked in liquids, chemical energy can come from dissolved molecules; if they are in the open, energy can come from light; if they are sitting in one place, they can be plugged into a socket; if they are moving around in the dark, they can run on batteries for a while, then run down and quit. Within these limits, much can be accomplished.

"Smart" is a relative term. Unless you want to assume that people learn a lot more about intelligence and programming, it is best to assume that these materials will follow simple rules, like those followed by parts of drawings on computer screens. In these drawings, a picture of a rectangle can be commanded to sprout handles at its corners; pulling a handle stretches or shrinks the rectangle without distorting its right-angle corners. An object made of smart matter could do likewise in the real world: a box could be stretched to a different size, then made rigid again; a door in a smart-material wall could have its position unlocked, its frame moved a pace to the left, and then be returned to normal use.

There seems little reason to make bits of smart matter independent, self-replicating, or toxic. With care, smart matter should be safer than what it replaces because it will be better controlled. Spray paint gets all over things and contains noxious solvents; the paperpaint described above doesn't. This will be a characteristic difference, if we exercise our usual vigilance to encourage the production of things that are safe and environmentally sound.

It may be fun to discuss wondrous new products, but they won't make much difference in the world if they are too expensive. Besides, many people today don't have decent food, clothes, and a roof over their heads, to say nothing of fancy "nanostuff."

Costs matter. There is more to life than material goods, but without material goods life is miserable and narrow. If goods are expensive, people strive for them; if goods are abundant, people can turn their attention elsewhere. Some of us like to think that we are above a concern for material goods, but this seems more common in the wealthy countries. Lowering manufacturing costs is a mundane concern, but so are feeding people, housing them, and building sewage systems to keep them from dying of cholera and hepatitis. For all these reasons, finding ways to bring down production costs is a worthy goal. For the poor, for the environment, and for the freeing of human potential, costs matter deeply. Let's take a closer look at the costs of molecular manufacturing.

Inflation produces the illusion that costs rise, when the real story is that the value of money is falling. In the short term, real costs usually don't change very quickly, and this can produce the illusion that costs are stable facts of nature, like the law of gravity or the laws of thermodynamics. In the real world, though, most costs have been falling by a crucial measure: the amount of human labor needed to make things. People can afford more and more, because their labor, supplemented by machines, can produce more and more. This change is dramatic measured on a scale of centuries, and equally dramatic across the gulf between Third World and developed countries. The rise from Third World to First World standards of living has raised income (dropped the cost of labor time) by more than a factor of ten. What can molecular manufacturing do?

Larger cost reductions have happened, most dramatically in computers. The cost of a computer of a given ability has fallen by roughly a factor of 10 every seven years since the 1940s. In total, this is a factor of a million. If automotive technologies had done likewise, a luxury car would now cost less than one cent. (Personal computer systems still cost hundreds of dollars both because they are far more powerful than the giant machines of the 1940s and because the cost of buying any useful computer system includes much more than just the cost of a bare computer chip.)

REAL ACCOUNT

Some costs apply to a kind of product, regardless of how many copies of it are made: these include design costs, technology licensing costs, regulatory approval costs, and the like. Other costs apply to each unit of a product: these include the costs of labor, energy, raw materials, production equipment, production sites, insurance, and waste disposal. The per-kind costs can become very low if production runs are large. If these costs stay high, it will be because people prefer new products for their new benefits, despite the cost-hardly cause for complaint.

The more basic and easier to analyze costs are per-unit costs. A picture to keep in mind here is of Desert Rose Industries, where molecular machinery does most of the work, and where products are made from parts that are ultimately made from simple chemical substances. Let's consider some cost components.

Energy: Manufacturing at the molecular scale need not use a lot of energy. Plants build billions of tons of highly patterned material every year using available solar energy. Molecular manufacturing can be efficient, in the sense that the energy needed to build a block of product should be comparable to the energy released in burning an equivalent mass of wood or coal. If this energy were supplied as electricity at today's costs, the energy cost of manufacturing would be something like a dollar per kilogram. We'll return to the cost of energy later.

Raw Materials: Molecular manufacturing won't need exotic materials as inputs. Plain bulk chemicals will suffice, and this means materials no more exotic than the fuels and feedstocks that are, for now, derived from petroleum and biomass-gasoline, methanol, ammonia, and hydrogen. These typically cost tens of cents per kilogram. If bizarre compounds are used, they can be made internally. Rare elements could be avoided, but might be useful in trace amounts. The total quantity of raw materials consumed will be smaller than in conventional manufacturing processes because less will be wasted.

Capital Equipment and Maintenance: As we saw in the Desert Rose scenario, molecular manufacturing can be used to build all of the equipment needed for molecular manufacturing. It seems that this equipment-everything from large vats to submicroscopic special-purpose assemblers-can be reasonably durable, lasting for months or years before being recycled and replaced. If the equipment were to cost dollars per kilogram, and produce many thousands of kilograms of product in its life, the cost of the equipment would add little to the cost of the product.

Waste Disposal: Today's manufacturing waste is dumped into the air, water, and landfills. There need be no such waste with molecular manufacturing. Excess materials of the kind now spewed into the environment could instead be completely recycled internally, or could emerge from the manufacturing process in pure form, ready for use in some other process. In an advanced process, the only wastes would be leftover atoms resulting from a bad mix of raw materials. Most of these leftover atoms would be ordinary minerals and simple gases like oxygen, the main "waste" from the molecular machinery of plants. Molecular manufacturing produces no new elements-if arsenic comes out, arsenic must have gone in, and the process isn't to blame for its existence. Any intrinsically toxic materials of this sort can at least be put in the safest form we can devise for disposal. One option would be to chemically bond it into a stable mineral and put it back where it came from.

Labor: Once a plant is operating, it should require little human labor (what people do with their time will change, unless factories are kept running as bizarre hobbies). Desert Rose Industries was run by two people, yet was described as producing large quantities of varied goods. The basic molecular-scale operations of manufacturing have to be automated, since they are too small for people to work on. The other operations are fairly simple and can be aided by equipment for handling materials and information.

Space: Even a manufacturing plant based on nanotechnology takes up room. It would, however, be more compact than familiar manufacturing plants, and could be built in some out-of-the-way place with inexpensive land. These costs should be small by today's standards.

Insurance: This cost will depend on the state of the law, but some comparisons can be made. Improved sensors and alarms could be made integral parts of products; these should lower fire and theft premiums. Product liability costs should be reduced by safer, more reliable products. Employee injury rates will be reduced by having less labor input. Still, the legal system in the United States has shown a disturbing tendency to block every new risk, however small, even when this forces people to keep suffering old risks, which are sometimes huge. (The supply of lifesaving vaccines has been threatened in just this way.) When this happens, we kill anonymous people in the name of safety. If this behavior raises insurance premiums in a perverse way, it could discourage a shift to safer manufacturing technologies. Since such costs can grow or shrink independent of the real world of engineering and human welfare, they are beyond our ability to estimate.

Sales, Distribution, Training: These costs will depend on the product: Is it as common as potatoes, and as simple to use? Or is it rare and complex, so that determining what you need, where to get it, and how to use it are the main problems? These service costs are real but can be distinguished from costs of the thing itself.

To summarize, molecular manufacturing should eventually lead to lower costs. The initial expense of developing the technology and specific products will be substantial, but the cost of production can be low. Energy costs (at present prices) and materials costs (ditto) would be significant, but not enormous. They were quoted on a per-kilogram basis, but nanotechnological products, being made of superior materials, will often weigh only a fraction of what familiar products do. (Ballast, were it needed, will be dirt-cheap.) Equipment costs, land costs, waste-disposal costs, and labor costs can be low by the very nature of the technology.

Costs of design, regulation, and insurance will depend strongly on human tastes and are beyond predicting. Basic products, like clothing and housing, can become inexpensive unless we do something to keep them costly. As the cost of improved safety falls, there will be less reason to accept unsafe products. Molecular manufacturing uses processes as controlled and efficient as the molecular processes in plants. Its products could be as inexpensive as potatoes. This may sound to good to be true (and there are downsides, as we'll discuss), but why shouldn't it be true? Shouldn't we expect large changes to come with the replacement of modern technology?

A CYCLE OF FALLING COST

The above estimate made a conservative assumption about future costs: that energy and materials will cost then what they do now, before molecular manufacturing has become available. They won't, because lower costs lead to lower costs.

Let's say that making one kilogram of product by molecular manufacturing requires one dollar for a kilogram of raw materials and four dollars for a generous forty kilowatt-hours of energy. These are typical present-day prices for materials and electrical energy. Assume, for the moment, that other costs are small. One of the resulting five-dollar-per-kilogram products can be solar cell paint suitable for applying to paved roads. A layer of paint a few millionths of a meter thick would cost about five cents per square meter to produce, and would generate enough energy to make another square meter of paint in less than a week, even allowing for nighttime and moderate cloud cover. The so-called energy payback time would thus be short.

Let's assume that this smart paint costs as much to spread and hook up as it does to make, and that we demand that it pay for itself in a single month, so we charge ten cents per square meter per month. At that rate, the cost of solar energy from resurfaced roads would be roughly $0.004 per kilowatt hour-less than a twentieth the energy cost assumed in the initial production-cost estimate. By itself, this makes the cost of production fall to a fraction of what it was before. Most of that remaining fraction consists of the cost of materials.

But the products of nanotechnology will mostly be made of carbon (if present expectations are any guide), and carbon dioxide is too abundant in the atmosphere these days. With energy so cheap, the atmosphere can be used as source of carbon (and of hydrogen, nitrogen, and oxygen). The price of carbon would be a few cents per kilogram-roughly a twentieth the original price assumed for raw materials.

But now, both energy and raw materials are a twentieth the original price, and so the products become cheaper, including the energy-producing products and the raw-material-producing (atmosphere-cleaning) products....

The above scenario is simple, but it seems realistic in its basic outlines: lower costs can lead to lower costs. How far this process can go is hard to estimate precisely, but it could go far indeed.

TOO CHEAP POWER

This argument will remind some readers of an old claim-that nuclear energy would lead to "power too cheap to meter." This assertion, attributed to the early nuclear era, has passed into folklore as a warning to be skeptical of technologists promising free goodies. Does the warning apply here?

Anyone claiming that something is free doesn't really understand economics. Using something always has a cost equal to the most valuable alternative use for the thing. Choosing one alternative sacrifices another, and that sacrifice is the cost. As economist Phillip K. Salin says, "There's no such thing as a free opportunity," since opportunities always cost (at least) time and attention. Nanotechnology will not mean free goodies.

But, one might argue, nuclear power hasn't even been inexpensive. If technologists could be so wrong back then, why believe a similar argument today? We are happy to report that the arguments aren't similar: any argument for "nuclear power too cheap to meter" had to be absurd even given the knowledge at the time, and our argument isn't.

Nuclear reactors boil water to make steam to turn turbines to turn generators to drive electrical power through power lines to transformers to local power lines to houses, factories, and so forth. The wildest optimist could never have claimed that nuclear power was a free source of anything more than heat, and a realist would have added in the cost of the reactor equipment, fuel, waste disposal, hazards, and the rest. Even our wild optimist would have had to include the cost of building the boiler, the turbines, the generators, the power lines, and the transformers, and the cost of maintenance on all these. These costs were known to be a major part of the cost of power, so free heat wouldn't have meant free power. Thus, the claim was absurd the day it was made-not merely in hindsight.

In the early 1960s, Alvin Weinberg, head of the Oak Ridge National Laboratory, was a strong advocate of nuclear power, and argued that it would provide "cheap energy." He was optimistic, but did his sums. First, he assumed that nuclear-power plants could be built a little more cheaply than coal-fired power plants of the same size. Then he assumed that the cost of fuel, waste disposal, operations, and maintenance for nuclear plants would be not much more than the cost of operations and maintenance alone for coal plants. Then he assumed that they might last for more than thirty years. Finally, he assumed that they would be publicly operated, tax free at low interest (which merely moves costs elsewhere) and that after thirty years the cost of the equipment would be written off (which is an accounting fiction). With all of that, he derived a power cost that "might be" as low as one half the cost of the cheapest coal-fired plant he mentions. He was clearly an optimist, but he didn't come close to arguing for power too cheap to meter.

BIBLIOGRAPHY

Attebery, B.: *Nanoculture: Implications of the New Technoscience*, Bristol, Intellect Books, 2004.

Bainbridge, W.S. : *Societal Implications of Nanoscience and Nanotechnology*, Kluwer, Dordrecht, 2001.

Berube, D.M.: *The Rhetoric of Nanotechnology*, Amsterdam: IOS Press, 2004.

Charles, P., and Frank J. Owens: *Introduction to Nanotechnology*, Hoboken, J. Wiley, 2003.

Credi, A.: *Molecular Devices and Machines: a Journey into the Nano World*, Weinheim, Wiley-VCH, 2003.

Drexler, K. Eric: *Engines of Creation: The Coming Era of Nanotechnology*, New York, Anchor Books, 1986.

Etzkowitz, H.: *Nano-Science and Society: Finding a Social Basis for Science Policy*, Dordrecht, Kluwer, 2001.

Frank J. Owens: *Introduction to Nanotechnology*, Hoboken, J. Wiley, 2003.

Glimell, H.: *Dynamics of the Emerging Field of Nanoscience*, Dordrecht, Kluwer, 2001.

Goodnick, M.: *Transport in Nanostructures*, Cambridge, Cambridge University Press, 1997.

Gregory, L.: *Nanotechnology*, New York, Springer, 1999.

Hayles, N.K. : *Nanoculture: Implications of the New Technoscience*, Bristol, Intellect Books, 2004.

Helvajian, Henry: *Microengineering Aerospace Systems*, El Segundo, CA: Aerospace Press, 1999.

Hessenbruch, A.: *Nanotechnology and the Negotiation of Novelty*, Amsterdam, IOS Press, 2004.

Jin Z. Zhang: *Self-assembled Nanostructures*, New York, Kluwer Academic/ Plenum, 2003.

Kurzweil, Ray: *The Age of Spiritual Machines*, New York, Penguin Books, 1999.

Lewis, J.: *Prospects in Nanotechnology: Toward Molecular Manufacturing,* New York, John Wiley & Sons, 1995.

Moravec, Hans: *Mind Children: The Future of Robot and Human Intelligence*, Cambridge, Harvard University Press, 1988.

Norio Taniguchi: *Nanotechnology: Integrated Processing Systems for Ultra-precision and Ultra-fine Products,* New York, Oxford University Press, 1996.

Owens, J.: *Introduction to Nanotechnology*, Hoboken, J. Wiley, 2003.

Postman, Neil: *Technopoly: The Surrender of Culture to Technology*, New York, Vintage Books, 1992.

Rietman, Ed.: *Molecular Engineering of Nanosystems,* New York, Springer, 2001.

Roobeek, A: *The Biotechnology Revolution?,* Oxford, Blackwell, 1995.

Shew, A.: *Discovering the Nanoscale*, Amsterdam: IOS Press, 2004.

Stephen M. Goodnick: *Transport in Nanostructures,* Cambridge, Cambridge University Press, 1997.

Tenner, E.: *Nanotechnology and Unintended Consequences*, , Dordrecht, Kluwer, 2001.

Vincenzo, M. Venturi, and A. Credi: *Molecular Devices and Machines: a Journey into the Nano World,* Weinheim, Wiley-VCH, 2003.

Weil, V.: *Ethical Issues in Nanotechnology*, Dordrecht, Kluwer, 2001.

INDEX

I

L

M

N

O

P

R

S

T

U

V

W

□□□